Intelligent Systems Reference Library

Volume 199

The aim of this series is to publish a Reference Library, including novel advances and developments in all aspects of Intelligent Systems in an easily accessible and well structured form. The series includes reference works, handbooks, compendia, textbooks, well-structured monographs, dictionaries, and encyclopedias. It contains well integrated knowledge and current information in the field of Intelligent Systems. The series covers the theory, applications, and design methods of Intelligent Systems. Virtually all disciplines such as engineering, computer science, avionics, business, e-commerce, environment, healthcare, physics and life science are included. The list of topics spans all the areas of modern intelligent systems such as: Ambient intelligence, Computational intelligence, Social intelligence, Computational neuroscience, Artificial life, Virtual society, Cognitive systems, DNA and immunity-based systems, e-Learning and teaching, Human-centred computing and Machine ethics, Intelligent control, Intelligent data analysis, Knowledge-based paradigms, Knowledge management, Intelligent agents, Intelligent decision making, Intelligent network security, Interactive entertainment, Learning paradigms, Recommender systems, Robotics and Mechatronics including human-machine teaming, Self-organizing and adaptive systems, Soft computing including Neural systems, Fuzzy systems, Evolutionary computing and the Fusion of these paradigms, Perception and Vision, Web intelligence and Multimedia.

Indexed by SCOPUS, DBLP, zbMATH, SCImago.

All books published in the series are submitted for consideration in Web of Science.

More information about this series at http://www.springer.com/series/8578

Rosalina Babo · Nilanjan Dey ·
Amira S. Ashour
Editors

Workgroups eAssessment: Planning, Implementing and Analysing Frameworks

Springer

Editors
Rosalina Babo
Porto Accounting and Business School
Polytechnic of Porto
Porto, Portugal

Amira S. Ashour
Department of Electronics and Electrical
Communications Engineering, Faculty
of Engineering
Tanta University
Tanta, Egypt

Nilanjan Dey
Department of Computer Science
and Engineering
JIS University
Kolkata, India

ISSN 1868-4394 ISSN 1868-4408 (electronic)
Intelligent Systems Reference Library
ISBN 978-981-15-9907-1 ISBN 978-981-15-9908-8 (eBook)
https://doi.org/10.1007/978-981-15-9908-8

This Springer imprint is published by the registered company Springer Nature Singapore Pte Ltd.
The registered company address is: 152 Beach Road, #21-01/04 Gateway East, Singapore 189721, Singapore

Preface

The use of collaborative work is increasing in various institutions. May they be higher education institutions, small or big companies, sporting organisations or other. This way of working can demand and improve a set of skills, such as cooperative ability, critical reasoning, creative thinking, responsibility, planning and communication. It also can provide motivation, learning opportunities and increase productivity. However, not every person has the same performance rate which leads to trust issues among the team members and a fault in the efficiency of the tasks. Hence, the assessment of the team members becomes an essential procedure, which can ensure that the individuals will meet the intended outcomes.

Each evaluator has their strategy to overcome issues and present viable solutions to distinguish team members and provide fair feedback. However, the way the individual contribution is assessed can be unclear and biased, and as a result, the evaluator has to resort to new methods of assessment. For that purpose, there are a variety of tools and techniques that can be used in several approaches.

This book seeks to overcome the assessment complexity using new frameworks and computer tools. It also reports the challenges and the new perspectives in developing e-assessment systems.

The book is a collection of nine chapters which are written by eminent professors, researchers and academic students from several countries. The chapters were initially peer-reviewed by the editorial board members who themselves span over different countries. The whole book is divided into two parts, namely Part I *Assessment and Collaboration* and Part II *E-Assessment Approaches*.

The first part, *Assessment and Collaboration*, comprises the following chapters and intends to provide a deeper understanding on the importance of assessment and collaboration, as well as some of the existing collaboration tools.

Chapter 1 by Katherina Gallardo tries to understand the importance of assessment literacy in the students' learning processes, by explaining the learning assessment practices according to psycho-pedagogical paradigms. Also, the significance of Information and Communication Technologies tools in the assessment is explained. The chapter makes some recommendations in the form of "lessons" to assist the educators in improving their assessment practices.

Chapter 2 by Salim Chujfi, Hanadi Traifeh, Thomas Staubitz, R. Refaie and C. Meinel analyses the collaboration in online training environments with the use of an online course and that the assessment of digital remote teams can be successfully implemented by means of assisting team building and encouraging virtual participation.

Chapter 3 by Paz Prendes-Espinosa, Isabel Gutiérrez-Porlán and Pedro A. García-Tudela presents an extensive literature on collaborative tools and their functionalities. It analyses the possibilities of collaboration in higher education with real examples.

Chapter 4 by Dalbert Oliveira and Ana Lúcia Terra explains in a more personal way the use of online collaborative tools, as well as their strengths and weaknesses. It uses the life history methodology to acquire information and carefully analyse the data that was gathered first hand.

Chapter 5 by Catalin Popescu and L. Avram provides an insight on online questionnaires and whether these are reliable to consider all the information related to project implementation. It intends to be a good practice guide in the implementation of projects to introduce new study programmes.

The second part, *E-Assessment Approaches*, composed by the following chapters, presents different approaches and tools to carry the assessment process.

Chapter 6 by Bastian Küppers and Ulrik Schroeder concerns the development of a framework to conduct evaluations in HEI with a Bring Your Own Device (BYOD) approach. The authors developed an application for electronic exams on students' devices in a BYOD environment: FLEX. FLEX's main goal is to be a software solution that enables electronic assessments within the MATSE educational program.

Chapter 7 by Joachim Maderer and Christian Gütl explains the need of a flexible and adaptive assessment system. This system should be able to: be integrated with different learning environments; recognize skills and the application of knowledge; and reuse learning and assessment items. Therefore, this chapter presents the Antares framework, its architecture, assessment interface, assignment and assessment engine, as well as an example of its working.

Chapter 8 by Paulo C. Oliveira, O. Constante, M. Alves and F. Pereira provides an interesting study in one Portuguese higher education institution, involving the use of electronic tools to assure that the students perform their preparatory work in advance to the laboratory classes.

Chapter 9 by Ursula Niederländer and Elisabeth Katzlinger provides a discussion on the use of a LMS Moodle plugin "StudentQuiz" to allow students to create questions and answers. This method also allows them to test, comment and rate the questions of their classmates.

The authors address important matters on assessment and e-assessment, collaboration environments and tools, as well as different and new assessment practices. Those interested in using new technologies and different learning environments will benefit from these studies. We hope that this book will assist researchers and students interested in carrying out further research in this area.

Porto, Portugal Rosalina Babo
Kolkata, India Nilanjan Dey
Tanta, Egypt Amira S. Ashour

Contents

About the Editors

Rosalina Babo is Coordinator Professor of Information Systems Department, School of Accounting and Administration of Polytechnic of Porto (ISCAP/IPP), Portugal. Since the year 2000, she is Head of the Information Systems Department and for about 12 years acts as a member of the university scientific board. Rosalina's international recognition was improved with the opportunity to be Visiting Professor at several universities in different countries, namely Belgium (KU LEUVEN), Croatia (University of Split), Kosovo (University of Prishtina), and Latvia (Latvia University of Agriculture). Rosalina was one of the founders of CEOS.PP (former CEISE/STI) research centre and its director for 5 years. Rosalina has served on committees for international conferences, and acts as a reviewer in scientific journals. As Book Editor, she collaborates with publishers such as Elsevier, Springer and IGI Global in the fields of data analyses in social networks and e-learning. Having several published papers, her main areas of research are e-learning, e-business, Internet applications focusing on usability and social networks.

Nilanjan Dey is an associate professor in the Department of Computer Science and Engineering, JIS University, Kolkata, India. He is a visiting fellow of the University of Reading, UK. Previously, he held an honorary position of Visiting Scientist at Global Biomedical Technologies Inc., CA, USA (2012–2015). He was awarded his PhD from Jadavpur University in 2015. He has authored/edited more than 70 books with Elsevier, Wiley, CRC Press, and Springer, and published more than 300 papers. He is the Editor-in-Chief of the International Journal of Ambient Computing and Intelligence (IGI Global), Associated Editor of IEEE Access, and International Journal of Information Technology (Springer). He is the Series Co-Editor of Springer Tracts in Nature-Inspired Computing (Springer), Series Co-Editor of Advances in Ubiquitous Sensing Applications for Healthcare (Elsevier), Series Editor of Computational Intelligence in Engineering Problem Solving and Intelligent Signal Processing and Data Analysis (CRC). His main research interests include medical imaging, machine learning, computer-aided diagnosis, data mining, etc. He is the Indian Ambassador of the International Federation for Information Processing—Young ICT Group and Senior member of IEEE.

Amira S. Ashour is Assistant Professor and Head of EEC Department, Faculty of Engineering, Tanta University, Egypt, since 2016. She is a member of the Research and Development Unit, Faculty of Engineering, Tanta University, Egypt, from August 2019. Ashour is Engineering Manager of Huawei ICT Academy, Tanta University, Egypt, from September 2019. She was Vice-Chair of Computer Engineering Department, Computers and Information Technology College (CIT), Taif University, KSA, for one year from 2015. She was Vice-Chair of Computer Science Department, CIT College, Taif University, KSA, for 5 years till 2015. Her research interests include biomedical engineering, image processing and analysis, medical imaging, computer-aided diagnosis, signal/image/video processing, machine learning, smart antenna, direction of arrival estimation, targets tracking, optimization and neutrosophic theory. She has 20 edited books and 4 authored books along with about 200 published journal and conferences papers. She is Series Co-Editor of Advances in Ubiquitous Sensing Applications for Healthcare series, Elsevier.

Part I
Assessment and Collaboration

Chapter 1
The Importance of Assessment Literacy: Formative and Summative Assessment Instruments and Techniques

Katherina Gallardo

Abstract Almost sixty years after Scriven and Bloom's accurate description and differentiation between summative and formative assessment, it would be quite vain to make sure that educators nowadays fully understand and use these two types of evaluation in their practice. Besides, it would be riskier to expect educators to use programs or design AI algorithms to make appropriate decisions to select and design instruments to make accurate judgments about learning and performance results without considering the difficulties in learning evaluation practices that have arisen in different educational contexts. The understanding of paradigms, educational models, and beliefs of educators around assessment practices constitutes mandatory tasks to consider as a point of departure in the era of ICT for learning purposes. Thus, the main objective of this chapter is to review the importance of evaluation literacy towards the complex challenge of planning, designing instruments, and interpreting results derived from learning assessment. Then, a reflection on the advances and difficulties found by researchers in different countries on formative and summative practices and results in higher education mainly is discussed. By the end of this chapter, some recommendations related to educator's training for improving and reinforcing conceptual and instrumental assessment practices are envisaged.

Keywords Assessment literacy · Formative assessment · Summative assessment · Higher education · Meta-evaluation

1.1 Introduction

Formative and summative assessment can be considered a worldwide educational topic based on the number of reports, scientific articles, and books published in the lasts three decades. On the one hand, UNESCO, OCDE, and World Bank as the

K. Gallardo (✉)
School of Humanities and Education, Tecnologico de Monterrey, Monterrey, México
e-mail: katherina.gallardo@tec.mx

R. Babo et al. (eds.), *Workgroups eAssessment: Planning, Implementing and Analysing Frameworks*, Intelligent Systems Reference Library 199,
https://doi.org/10.1007/978-981-15-9908-8_1

more relevant international organizations associated to educational matters around the world have established totally or partially global as well as local recommendations to profit these two types of assessment as powerful ways to improve learning [56, 76, 82].

On the other hand, also educational researchers are interested in studying formative and summative assessments in daily classroom activities as it has been proved that both types in different ways influence students' academic progress [47, 54, 84]. Besides, formative and summative assessment becomes a more complex topic when variables such as teacher experience, type of contents, and social and cultural contexts are involved in the understanding of assessment planning and results [53].

Moreover, in several learning scenarios, formative and summative assessment are now supported by ICT tools. Educators use different kinds of information and communication technology (ICT) applications to plan, design instruments, design activities, analyze results, grade, and give feedback to students [83]. The options of electronic tools for assessment tasks have increased exponentially in the last five years. Undoubtedly, new and more natural ways to conduct formative and summative assessment will keep rising in the next years supported by AI and machine learning [1, 36].

Even if new and useful knowledge on formative and summative assessment as well as innovative ICT assessment applications are available for improving these particular educational tasks, it can also be affirmed that there are still many difficulties to solve around classroom assessment practices. These difficulties are mainly related to assessment literacy and educators' beliefs about assessment [4, 75].

The objective of this chapter is to get a deeper understanding of how vital assessment literacy is. Then, it is relevant to reflect upon this topic seen within the complexity of the educational, social, and cultural environment. These reflections will permit to give some punctual recommendations around formative and summative assessment practice for higher education mainly, regarding a near future where ICT use for learning assessment becomes foreseeable.

1.2 Paradigms of Education and Educational Model

It is considered appropriate to begin this section by stating some clue ideas about the importance of psychopedagogy paradigms as well as the influence educational models must have on learning assessment. On the one hand, a paradigm can be defined as a mental representation of how an item or an idea is structured and how it functions in a specific context and time. The legacy of the last century on the development of psychological approaches related to human learning has left a considerable inheritance. Thus, five psycho-pedagogical paradigms arose: Behavioral, Humanist, Cognitive, Sociocultural and Constructivism paradigms have been considered for years as role models in the teaching–learning process [16, 35].

On the other hand, an educational model is the application of educational paradigms that an institution assumes and serve as a reference for the functions it fulfils: teaching, research, and establish links with other institutions, to carry out an educational project. Then, an educational model is based on the institutional history, values, vision, mission, philosophy, objectives, and formative purposes [78]. Indeed, the learning model is a conceptual framework that describes a systematic procedure in organizing learning experiences to achieve specific goals and serves as a guide for implementing the learning activities. Undoubtedly both, the psycho-pedagogical scope, as well as the institutional orientation, give meaning to educational practice. Then, the essence of the educational process influences the approach educators takes to plan, execute activities, and assess learning outcomes [37].

These statements are useful to understand that learning assessment not only refers to techniques or procedures in isolation [66]. The way an institution explains the essence of its educational orientation and model influences directly in the main decisions related to the establishment of learning and teaching strategies, methods, techniques, and practices. Then, the point of departure for educators should be these questions: what decisions am I taking to conform an assessment system in concordance to paradigms and educational model of the institution? Is my assessment practice in agreement with the institutional, educational goals? Unfortunately, these questions are rarely asked.

Apparently, in most of the cases, it is not well understood that paradigms in various ways guide relevant formative decisions. For instance, from the Behaviorist paradigm, it is understood that the learning process is related to accomplish specific learning tasks and that the level of accomplishment is represented on quantitative results and get meaning into a scale. The educational model based on objectives is closely related to this learning scope. Its statements configure the teaching and learning tasks. Then, learning assessment is based on practices to collect specific information on the acquisition of certain knowledge that permits either success in reaching a certain level of learning and achievement or failure.

On the opposite side, from the Cognitive paradigm, it is expected that learning process is focused on thinking processes, going from simple to complex that permit learner to identify, comprehend, analyze and use knowledge for solving problems [16]. The educational model based on thinking development emphasizes the achievement of autonomous reasoning. In this direction, learning assessment should be a focus on a variety of activities and mechanisms to promote the understanding of the own learning style, thinking process, and appropriation of knowledge. Then, assessment results get different meanings to the learner, going beyond a numerical scale of success or failure.

In the understanding of the author, statements around how psycho-pedagogical paradigms and educational models configure the educational intention, this point should be considered the cornerstone of the principles that rule assessment decision-taking in the classroom. To the extent that educators consider these psycho-pedagogical fundamental elements into their practice, taking decisions around assessment would be more coherent and aligned to educational goals.

1.3 Formative and Summative Assessment

Learning assessment was classified as formative and summative assessment six decades ago. Since Scriven's masterpiece published in 1967, several explanations and discussions have emerged around this classification up to now [65]. The level of relevance of the understanding of what these two types of assessment require and offer to students and educators is considered high.

In this section, readers can find a brief appreciation of the evolution of these concepts and the actual meaning of them. From these two types of assessment, it could be affirmed that formative assessment has been a more debated topic than summative assessment, not only from the psycho-pedagogical view but from many other disciplines.

The first definition of formative assessment given by Scriven was referring to provide some data that permit successive adaptations of a new program during the phases of development and implementation [65]. A few years later, the concept of formative assessment was adopted by Benjamin Bloom to enhance his definition of mastery learning [12]. A first feature that defined formative assessment was its role for closing learning goals gaps between the actual level of the work and the required standard. Then, specific corrective activities could be designed and done to correct learning difficulties [25].

In the '70s, Bloom's understanding of the potential of formative assessment was also enriched by practical recommendations. His initial proposal follows the establishment of some strategies that educators could do during instruction, such as: use formative assessment after each lecture, design and apply formative quizzes, give feedback and correction to each student as a way to reach remediation, assess the level of achieved standards in parallel to all learning objectives, assign a positive perspective to sanitation as its benefits towards improving students' achievement.

The decade of the '80s was crucial for getting to a better understanding of the possibilities that formative assessment gives to students and educators. Thus, it was found that useful feedback as possible as long as students develop their capability to judge the quality of their work [59, 60]. This statement guided educators to reinforce and renew some instructional practices as well as assessment techniques such as establishing standards students may refer to the beginning of the learning process for making judgments of their progress and work autonomously.

During the '90s, some other related variables and effects were studied around formative assessment practices beyond cognitive aspects. Then, self-assessment, motivation for learning, engagement, as well as communication around feedback considering the relationships among students and teachers as well as the social context, were studied as factors that could affect formative assessment results. Besides, findings around specific needs and complementary skills educators have to learn and develop for conducting this type of assessment were discussed [53, 55]. Some emergent concerns went around the lack of pedagogical preparation and practice for doing formative assessment practices in the classroom, specifically in tertiary education.

In the first decade of the new millennium, the number of studies about formative assessment specifically increased a thousand per cent concerning last decade scientific production [64]. Indeed, critical new elements were found and discussed. In the first place, it can be considered trustworthy that theory about formative assessment had been proposed [10]. Necessary elucidations were made about the types and moments of interaction with the teachers, learners, and the subject discipline; the teacher's role towards the regulation of learning; the feedback and the student–teacher interaction to get to an understanding about strengths and gaps for reaching learning goals; and the student's role in the learning process, as peer and as owner of the learning process.

In addition to this new theoretical input, it also was published a feedback model from a meta-analysis [28]. According to the authors, this feedback model was possible after working with 500 meta-analyses, involving 450,000 effect sizes from 180,000 studies, representing approximately 20–30 million students, on various influences on student achievement. This effort permitted to conclude that feedback has almost the same impact on learning than instruction. Besides, the model enhanced the role of students as the center of the knowledge and reflecting process. Moreover, the model indicates the position of educators as a guide and support along any learning path. Nevertheless, some other studies of this decade were not precisely aligned to encounter more explanations of the benefits of formative assessment for learning. In this decade, some studies were conducted to criticize and questioned the power of formative assessment [18, 57]. The main counterpart ideas were focused on weak theoretical foundations and lack of reliable results about the positive effects of formative assessment in the classroom. Specific suggestions that arose from these studies were related to the use of appropriate methodologies, and statistical techniques to make possible real improvement of instructional practices.

In addition to the publication of profound studies around formative assessment theoretical statements and several applications in the instructional process, the design of different ICT solutions considering correspondence to international formative assessment standards was also a characteristic of this decade. In this concern, specific developments such as the Classroom response systems supported on Technology-enhanced formative assessment were applied and studied. This technology was proved to use sets of questions working together to target instructional objectives on science education purposes [8, 83]. Some other applications like Alice [32] were design for improving data collection, reducing the loss of data, and improving the quality of the assessment instruments obtained from formative assessment practices to benefit results of summative assessment in terms of validity.

In the second decade of the new millennium (from 2010 up to now) the number of published articles, proceedings, and books about formative assessment has tripled the number of publications corresponding to the period 2000–2009. From all these contributions, three topics are considered relevant for strengthening conceptual and practical elements of formative assessment. The first one is the continuous revision of the theoretical and practical aspects, going further face to face modality [9, 38]. The second relevant topic is the design and use of holistic and analytic rubrics as a way to respond to the formative assessment principle of establishing standards and criteria from the beginning of the learning process and make clear the expected performance level for complex tasks [26, 49]. The third relevant topic is about the incursion of new technological development for introducing 2D and 3D consoles in the classroom as well as online games. These resources open a broad kind of possibilities for amusingly conducting formative assessment for millennials and Z generation students [30, 77].

In the side of summative assessment, the first definition expressed that this type of process serves to understand if the object being evaluated (program, intervention, and person) met the stated goals. Years later, some other definitions enriched this first appreciation, giving a more detailed meaning determining summative assessment as the judgement which encapsulates all the evidence up to a given point. This point is seen as a finality at the end of the analysis [73].

Summative assessment, contrary to formative assessment, has experimented with different pace of publications along the last six decades. The number of products around this topic has reached just 10% in comparison to formative assessment in the same period. The most related items allied to summative assessment in these publications are reliability, validity, test design, teaching, scoring system, and accreditation [64].

Curiously, in the most cited article of summative assessment along these almost 60 years [39] there are quite important issues that point out several problems found in the educational practice in different directions: conceptualization, instrument design, and establishment of judgments from the results. Three outstanding reflections found in this article are:

- The relevance of accuracy and reliability as quality test factors in most of the cases oblige educators to turn into pieces of curriculum content that had been learned not in isolation but in interrelation to other elements in authentic situations. In the words of the author, here there is a typical case of the juxtaposition of engagement and quality.
- The establishment of criteria derived from learning goals, a process that can take a lot of time and effort, especially for complex skills. In most of the cases, after testing students and giving a grade, it is impossible to know what criteria have been used or what meanings had been attached to them; Remove any existing section breaks.
- The misunderstanding of grading in local, regional, and global contexts. On the one hand, a grade opens a silent gap that makes the learning processes involved invisible. On the other, the context factor (learning conditions, and abilities of the test designer) makes it impossible to rely on a final grade as a warrant of learning. In this third point, assessment validity is reluctant.

Finally, there is a convergent point that must also be understood. Assessment in the classroom goes far beyond a confrontation between formative and summative assessment practices. The integration of both types of assessment, as Scriven affirmed, needs to be done for constructing a reliable assessment system. In the literature, several studies highlight that the coexistence of formative and summative in harmony along the learning process benefits students' academic performance and achievement as well as give valuable information to educators for improving instruction [24, 31, 51]. For this reason, some technical and instrumental aspects related to both, formative and summative assessments are addressed in the next sections of this chapter.

1.4 Assessment Literacy

Educators impact students' learning process every day. Several factors could make the learning process a profitable experience to obtain positive effects: the application of specific teaching methods, the selection of the didactic and reading materials as well as the decisions educators might take on learning assessment. Then, knowledge, skillss and understanding around learning assessment decisions could be defined as the complex task of assessment literacy. Nevertheless, this definition stays quite simple in comparison to what preparation for executing learning assessment processes implies.

The work of assessment literacy in this section of the chapter is focused on its relevance. It is considered a concept that goes beyond school frontiers. Thus, assessment literacy integrates not only theoretical but personal, institutionals and social elements in its complexity. Figure 1.1 contains the items that conform to a proposal for assessment literacy model that, according to the author, integrates the crucial factors for its understanding and practice.

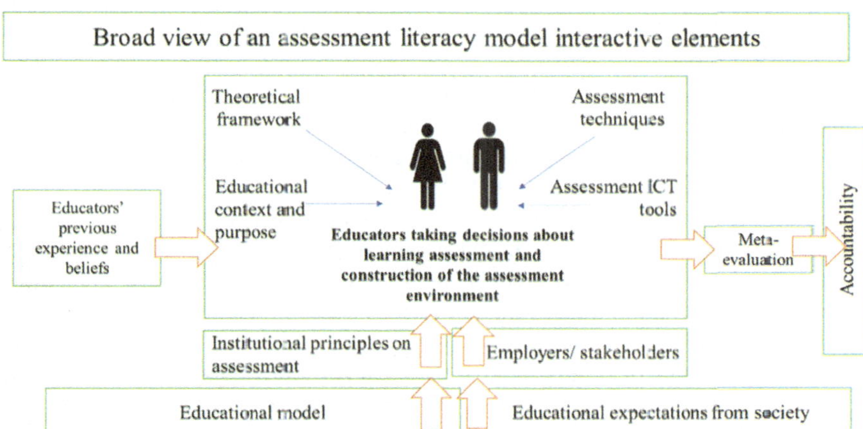

Fig. 1.1 Assessment literacy model: integration and interaction of main elements

1.4.1 Beliefs and Decision Making for Learning Assessment

Educators' beliefs as a research subject of interest and its influence on students' achievement and performance have taken great importance in the last years as part of the elements that conform to the assessment literacy task. Educators' beliefs could be defined as a set of conceptual representations which store general knowledge of objects, people and events, and their unique relationships [29]. Several studies in the last decade have indicated that studies on educator's beliefs are especially useful to understand teachers' perspective and practices related to their goals and actions in their professional practice [21, 63].

An excellent way to understand the relevance of this topic might be through the following statement that contains some meaning about beliefs and the connection to teaching practices: If an educator believes that teaching mainly consists in the transmission of knowledge, then students' role could be circumscribed to the passive accumulation of information. Nevertheless, if this belief changes and transforms educators' approach of teaching into a co-constructive process that involves teachers and students to reach a goal, then; as a result, a higher possibility of transforming educational practices could happen. This new belief would open up a new understanding of teaching as a way to support students' learning to stimulate their engagement and interest in learning in specific topics with practical ways to do so instead of passive.

The statement above can also be understood from the framework of assessment practices. Indeed, some of these educator's beliefs studies integrate exciting facts about educators' decisions around learning assessment [7, 14, 23, 80]. Relevant aspects could be summarized in this point to fully understand the relevance of teachers' beliefs in assessment.

- Few studies have explicitly examined teachers' beliefs up to now in relation to learning assessment process. There is a need to continue making research about influences of teachers' past and current experience on design and analysis of assessment in the classroom [7, 14].
- There is a need for giving relevance to academicians' beliefs about learning to pursue meaningfulness in teacher training programs [7]. The inconsistencies found from beliefs to practice help identify relevant aspects to incorporate in teachers' training programs or to reconsider in institutional norms about assessment [80].
- The understanding of beliefs about assessment permit to identify if either internal or external factors are guiding the learning assessment process when analyzing educators' daily practice [14, 23].

1.4.2 Theoretical Approaches Related to Learning Assessment Practices

Assessment literacy is a complex task that involves theories from the psycho-pedagogical approach that sustains its practices. As it was explained before, the learning paradigm, as well as the educational model, are clue elements that guide formative ideologies and actions [16, 78]. Then, these theoretical elements influence directly the way educators understand and decide how to conduct assessment procedures, accompanied by their previous experience and beliefs.

On the one hand, theoretical approaches from Behaviorism, Cognitivism, Constructivism, Humanism, and Sociocultural or Situated approach indicate specific features that the educational process must contain. This same phenomenon happens around learning assessment and configures the way to conduct formative as well as summative assessment decisions in the classroom. In Table 1.1, some foundations of learning and assessment practice are displayed [34, 35].

On the other hand, there is a branch of complementary psycho-pedagogical theoretical approaches that support explanations about not only how human beings are capable of learning contents, but also movements as well as manage social and emotional elements in interaction with others. These referential frameworks are known as learning taxonomies. Learning taxonomies feature a well-defined hierarchy of categories that attempts to capture the spectrum of learning processes: These taxonomies are helpful tools educators use for planning, instrumenting, and implementing assessment techniques. This is the reason why taxonomies also constitute a relevant aspect to consider in assessment literacy.

In literature, Bloom's learning taxonomy [11] could be considered one of the most popular of the last century. Bloom and his colleagues developed an interesting proposal for understanding the way human dimensions such as cognition, psychomotor, and emotional -affective domains can be distinguished from the instructional aspect. Bloom's masterpiece has been revised and counts on enriched contents [5]. Nevertheless, other proposals are as quite remarkable as Bloom's. Marzano and Kendall's New Taxonomy [43] offers a different point of view to understand the connections between contents and cognitive procedures. The authors have also implemented a proposal for the psychomotor domain. Besides, there is another proposal: Experiential Domain taxonomy [48] based on the understanding of experience as a hierarchy of stimuli, interaction, activity, and response with exposure and culminating in the dissemination.

Even though the intention of this chapter is not to give a profound resemble of the use of each learning taxonomies, it is highly recommended to consider that educators need to learn and apply them as a crucial theoretical element. There are specific benefits, while taxonomies are regarded in the assessment process [33]:

- Educators need to refer to a common framework to understand the learning process and make the individual as well as collective decisions about progression and actions.

Table 1.1 Foundations of learning and assessment practice

Theoretical approach	Learning foundations that configure learning assessment perspectives
Behaviorism	Learning is a conditioned response to external stimuli Rewards, withholding, and punishment are the most used ways of forming or extinguishing habits Learning can best be accomplished when elaborate performances are decomposed into parts Each element that conforms to a complex learning process should be practiced and reinforced Only observable behaviors are valid elements to give a judge about the sufficiency of learning Achievement is often equated with the accumulation of skills and the memorization of information in a given domain, that allows the learner to provide a rapid answer and demonstrate accurate performance
Cognitivism	Learning is determined by what people think and need Learning requires the active engagement of learners There is an emphasis on understanding as a way to reach learning goals Educators' primary role is to help novice students to acquire expert knowledge of conceptual structures and processing strategies Problem-solving is seen as a didactic means for knowledge construction Deductive and inductive reasoning are essential as evidence of analytic thought
Constructivism	Prior knowledge is a powerful determinant of a students' capacity to learn new things Learning principles go around how people construct meaning and make sense of the world through structures and concepts Construction of knowledge and meaning happens in community. Students work primarily in groups
Sociocultural or situated	The constant interaction between actions alters the context. Then, the context changes thinking Learning is a social and collaborative activity. People develop their thinking together. Then, conforming learning communities is part of the learning process Knowledge of not abstracted from context but seen in relation to it. Then, it is difficult to judge if an individual reached the learning goals from decontextualized situations
Humanism	The learning process implies the activation of cognition, emotion, interests, motivation, and potential of students The understanding of students' inner thought makes clear their difference in interests, needs, and experience while learning Good teacher-student interaction is mainly considered for promoting positively in constructing the learning environment Teachers reflect on their teaching style and attitude to understand themselves as educators and improve their acts continually

- Common frameworks in the different domains (cognitive, psychomotor, or affective-social) could give clear and justifiable reasons to link objectives, assessment, and outcomes, with appropriate teaching and assessment methods.
- Educators, as well as students, could revise learning progress in the different levels of the domain according to expected performance.

1.4.3 Assessment Techniques: Suggested Procedures to Make Decisions

In this chapter, assessment techniques refer to all the possible ways educators can take to plan and design tasks or instruments for collecting information that reflect students' performance. At this point, it is essential to clarify that these techniques are not intended and applied in isolation. Indeed, this becomes part of the classroom assessment environment where assessment purposes, tasks, performance criteria, and standards derive in the production of learning outcomes and later statement of feedback [13]. This environment is a product of assessment choices connected to the theoretical elements, the educational model, and the institutional operational features: format, frequency, and instructional functions.

Interminable discussions could be included in this section about where to start designing an assessment environment as well as which and how many assessment techniques should be included. Assessment literacy studies show several standpoints on how to do so [31, 40, 45, 61]. Figure 1.2 has been designed for explaining a suggested way based on assessment literacy literature. This proposal is divided into

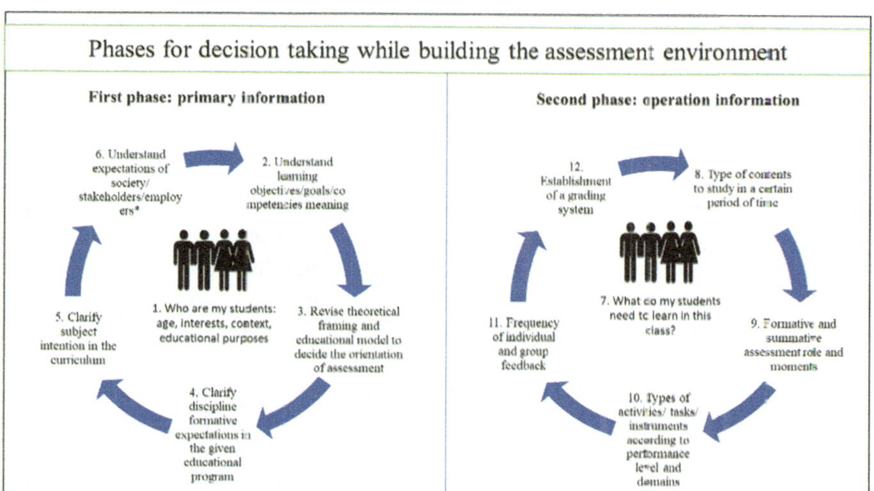

Fig. 1.2 Phases for learning assessment decision taken by educators

two phases for primary and secondary decisions while building or improving the assessment environment.

The first phase of the process permits educators to reflect upon different general aspects. It would be possible that the answer to these first six items permit educators to establish stronger ideas about the relations among students' characteristics as learners, institutional, educational model and expectations about the educational process, identify where the subject matter is located concerning others. Besides, society, stakeholders or employers needs and expectations about educational outcomes are also part of this complex reflection. It is also recommendable to revise materials containing information about opinions and hope of society in general and employers in particular, especially in the case of higher education scenarios.

The second phase goes directly to the assessment process to be planned and followed in the classroom. Steps 7 and 8 permit educators to specify learning needs in a target population. Steps 9 and 10 are directly connected to the theoretical approach and learning foundations (see Table 1.1) as well as with the learning taxonomies selected for organizing and systematizing learning progression. At these points, educators need to establish the intention of assessment according to objectives, goals, and content for then determining when and how the assessment process will take place. Finally, with steps 11 and 12 educators can end up the assessment cycle by deciding the frequency of feedback, the way it is going to be delivered and the organization, and the grading aspect according to the institutional principles.

As can be inferred from all the above, the construction of the assessment environment is a complex task that demands reflection, sensibility, and preparation from educators. Assessment literacy principles are dynamic contents and processes that required constant interaction, thinking, and decisions among educators, directors, and social agents, all interested in pursuing an assessment process that benefits students to reach their learning goals.

1.4.4 Workgroups Assessment

Collaboration has become a relevant competence to be developed along with higher education. It is considered a valuable soft skill for further professional development as well as a powerful indicator of employability [41]. Collaborative work is defined as students working together in groups within a physical or virtual environment towards defined learning purposes and goals. It could be done with some or no tutor surveyance [15].

The success or failure of collaborative work depends on many different variables such as the number of students in a group, the possibility or not for students to choose group members, rules establishment, frequency of group meetings, and sense of responsibility [9]. Nevertheless, one of the main elements for collaboration to become a successful means for learning is related to assessment decisions and practices. At this point, educators need to select the grade in which students will have certain participation while executing both, self and peer assessment along with their collaboration experience as well as how this participation will configure feedback and grades. Steps 9–12 (Fig. 1.2) are referred to organize the process and explore the possibilities educators have to choose assessment practices for conducting collaborative work.

Step 9 makes it possible to make some initial decisions: how to explain the intention of each type of assessment, formative or summative, and the role students will have in each one. First, it is important to make clear for students the difference between both types of assessment: formative refers to ways to reaching learning goals working on the actual level of the work and the required standard, while summative permit students to count on a holistic type of judgment which encapsulates the learning evidence up to a given point [74]. Then, educators need to clarify from the beginning of how formative and summative assessment methods will be applied. From this point, the expected level of performance should be established. If performance integrates different dimensions as cognitive, psychomotor, and social-affective elements, criteria for assessing each dimension should be explained as well as the ways to qualify them.

Step 10 refers to the type of assessment activities. Educators are asked to establish at least three elements:

- The timetable for organizing collaborative work
- The inclusion and use of self-assessment and peer-assessment within formative and summative purposes
- The way self and peer assessment results affect or configure feedback and grading. These assessment activities should be carefully chosen. In the case of self-assessment, at least four different scopes could be selected [48, 67]: self-grading, self-rating by using certain criteria established by the educator in a rubric, self-rating by using students' criteria and standard self-assessment and learning contract design, applied only if students are also asked to decide contents as well as activities to get to a certain learning goal. In the case of peer assessment, at least three varieties could be used: peer feedbacking and peer grading with the option of personalizing each group member grade by making an algorithm for obtaining an index calculation (from 0 to 1) to multiply final group grade or add points to the final result. Besides, educators will decide if peer assessment will have to be managed in quantitive and qualitative way and if anonymously [46, 69].

Step 11 refers to the need of establishing dates and ways to participate in self-assessment and peer assessment. It is important to clarify specific dates as well as the use of forms, surveys, questions, and scales they will be using for handing these types of information.

Step 12 refers to establish how partial and final grades are configuring final results for group and individuals. All the decisions taken in step 10 should be well-structured at this point for avoiding confusion or argues along with the collaborative work.

1.4.5 ICT Tools for Assessment

For decades, learning assessment has also been an educational phenomenon of interest for computing engineers. E-assessment denotes end-to-end electronic assessment processes where ICT is used for the presentation of assessment activity, and the recording of responses. Interdisciplinary efforts in this matter have been designed for solving general as well as specific problems while conducting the formative or summative assessment. As a result, nowadays, educators can count on numerous possibilities of electronic assessment tools. It can be affirmed that these tools have been developed mainly to attend to the need for conforming question databases to design and apply quizzes and tests. Strategies related to strengthening the expertise of students responding quizzes to improve final summative tests results have been widely considered [17, 44].

Significant benefits of ICT tools for assessment have contributed to the fast development of these solutions: fewer hours devoted to testing design, lower administrative costs in the reproduction and application of tests, automatic, accurate statistical analysis of items and for obtaining more consistent tests, and more possibilities of sharing database and results for different publics. These are just a few of the advantages of the integration of these computing solutions to assessment environments [44].

Besides, the increasing interest in the online education modality has potentiated the need in the development of more reliable assessment tools [22] that count on some specific requirements as usability, accessibility, and interactivity for planning and conducting learning assessment. Recent research offers a classification of assessment tools typically used in online and blended learning modalities after a systematic literature review [20]. As a result, it is stated that educators can count today on manual assessment tools, semi-automatic assessment tools, and automatic assessment tools. The second and third options are most famous as the main objective of using these tools is to support immediate feedback. This action implies the delivery of results after the assessment process in the shortest period.

One distinctive of the evolution of these tools is the insertion of new concepts to enrich formative assessment intention. The practice of just applying questions randomly, collect information about students' score, and give automatic feedback have evolved. There are new proposals that integrate gamification as a motivating way to reach learning goals [79, 85]. There are also new alternatives that stimulate smiling while responding to formative assessment activities as a way to elevate motivation and satisfaction levels to maintain learning engagement [81].

Electronic tools for designing rubrics and compile portfolios constitute another critical branch of application development that have helped educators to create and establish both, criteria and performance level for formative or summative assessment intentions [2, 67]. Nowadays, it is quite common to find this type of tool available as an independent open-access application or as part of LMS assessment functions. The advantages of design rubrics using electronic applications are vast:

- High possibilities of working in teams rather than in isolation for the establishment of relevant criteria and expected performance.

- Run a statistical analysis for obtaining inter-rater and intra-rater reliability.
- Lower time-consuming while design and use rubrics.
- Automatic possibilities to convert scales to scores.

New generation virtual assessment tools are also a quite important topic nowadays [3, 58]. Simulators used inside the classroom based on different technologies as virtual reality augmented reality, or 360° video in preparation for authentic professional exercise scenarios have developed integrated assessment process to give automatic feedback to students while the practice has been concluded. According to the results of these studies, there is no plausible difference between the quality of feedback and academic achievement using or not these virtual assessment tools. The difference is focused on another type of element such as motivation, engagement, and possibilities of conducting self-assessment frequently.

All these new alternatives for planning, designing applying, and giving results including self-assessment and peer assessment information into grades are undoubtedly valuable possibilities for educators in different teaching contexts. Nevertheless, the risk of just using these varied and multiple tools without previous analysis of learning goals, students' needs, and formative and summative roles in the learning process is latent. The concept of conforming to a useful and meaningful assessment environment could be left aside if the selection of electronic tools become the fundamental aspects of educators while taking assessment practice choices. Then, assessment literacy becomes an important topic to help educators reflect and support their decisions.

1.4.6 Meta-Evaluation and Accountability

Other clue elements in assessment literacy must be considered in educators' training process that directly affects education quality matters. These elements are assessment meta-evaluation and accountability. Meta-evaluation, on the one hand, is process is carried out based on the documentation of the evaluation processes and results, which must be done in a wide-ranging manner to establish criteria around the fulfilment of learning. Thus, it helps to delineate, obtain and apply judgments about the usefulness, feasibility, and accuracy of the requirements, instruments, and activities that conform to the assessment environment [71, 72]. The meta-evaluation process should be a systematic and constant practice in any educational scenario. There are important topics that must be considered when conducting a meta-analysis on learning assessment:

- Appropriate level of relevance of the design and application of activities, projects, quizzes, and tests according to partial and global results obtained framed by objectives or competencies.
- Level of educator's impact on students learning and assessment. This data comes from the voice of students that are registered to estimate the instructor's effect on learning and assessment.

- Utility of the assessment concerning the assessment purposes and information needs of students and educators. Thus, assessment should have been informative, compelling, and applied at an appropriate time.
- Feasibility of the assessment concerning a system based on a plan and adapted to the context and learning circumstances, ensuring cost-effective results.
- Accuracy of the assessment as it must provide appropriate, valid, and applicable information that is developed using conceptually and methodologically robust tools.
- Legally and ethically correct assessment procedures as it is known that not well-conducted assessment procedures could derive in the affection of people and organizations in different ways. Then, assessment should have been conducted about ethical and legal issues.

On the other hand, accountability is the process of informing parents, stakeholders, and society about how well a school is reaching the expected educational quality level as well as the quality of the social and learning environment [42]. Even if this topic is relevant for all educational levels, it is crucial for tertiary education. In terms of funding and return of investment, elements such as transparency and clear prescriptive guidelines are essential to driving evaluation and quality control. Thus, the quality of conducting learning assessment and improving the assessment process through meta-evaluation affects directly the accuracy of information related to students' academic achievement.

1.5 Recommendations Related to Educators Training for Improving and Reinforcing Conceptual and Instrumental Assessment Practices

The prescriptive nature of this section takes the author to reunite some basic advice to impulse a culture of high-quality learning assessment practices. Besides, some complementary specific recommendations are given for higher education educators. All these suggestions are based on studies that have left "lessons learned" about assessment procedures in different educational communities along the time. A total of four lessons learned are described below.

The first lesson is allowing educators to base their assessment decisions on assessment literacy principles and support collegiate work base on their experiences. This process takes time as elements like previous experience and beliefs about their role as evaluators are inserted in such a way that could permit or interrupt the appropriation of principles, concepts, and skills to conduct assessment [6]. For higher education, the specific advice at this point is asking support from the educational faculty or the evaluation center to get some guidance or model on relevant decisions around assessment practices [19].

The second lesson is the coexistence and harmony that educators need to build between formative and summative assessment in the learning environment [24, 31,

51]. The equilibrium between these two approaches need to end up in a coherent system, in which neither formative nor summative assessment is the critical part, but an understandable way to impulse students to reach their learning goals and interests. In the case of higher education, one common practice is to privilege summative assessment to respond to the requirement that a certification process could demand. Although this is great responsibility for educators, it is also confirmed that students perceive great learning support when receiving feedback for improving their performance [28]. Thus, formative assessment is a desirable practice even for learning activities with a more summative propensity.

The third lesson addresses the issue of selection and use of electronic and virtual assessment tools. Factors such as modality, time, effort, and number of students are, most of the time, the main reasons for selecting one or a variety of electronic assessment tools. Nevertheless, these factors are totally on the side of the operational information (see Fig. 1.2). The recommendation for educators in this decision making procedure is, first, understand the essence of learning goals and the theoretical framework (paradigm, educational model) to take decisions on the assessment needs to connect students and favor learning engagement and commitment [62]. Whenever this is clear, then the process of decisions taking corresponding to the e-assessment tools for fortifying the assessment environment could provide an excellent service to do so. It is not recommended to do this on the opposite side.

The fourth lesson is taking the time and try to conduct a meta-evaluation process to improve educators' practice as well as help target public to interpret learning results in good and accurate ways. This process should be made as clear as possible to enhance the objective of communication and transparency inner and outer institution [42]. In higher education, all faculty need to work together towards an improvement-oriented assessment system for ensuring excellence in education. It can be affirmed that this is one of the most relevant clues for real educational reforming on colleges and universities.

1.6 Discussion

Conducting the learning assessment process is a matter of taking decisions. These decisions derive from different inner and outer sources: educators' personal beliefs and experiences about assessment, educational paradigms, educational models, curricular contents, expected performance level, ethical principles, and degree of students' participation in the process. Thus, lack of opportunities to get deeper into assessment literacy, conceptual and instrumental misunderstanding as well as dismiss learning assessment results for improving teaching quality are some of the big mistakes institutions and educators do repeatedly [27, 35, 40, 50].

It could be affirmed that formative and summative assessments are certainly misunderstanding conceptual problems that have been the center of debate for years [10, 18, 74]. Giving attention to the intention of assessment rather than to mechanisms has been most of the time a critical issue. Nevertheless, the revision of the literature around this topic is permitted to understand that assessment intentions should be placed at the beginning of the planning process.

Formative and summative assessment intentions coexist in the same learning environment and make it possible to inform students and educators about students' progress and compliment of goals Afterward, decisions about mechanisms should be taken. Quizzes, summaries, essays, conceptual maps, projects, tests, and surveys are ways to collect information for making inferences about students' learning progress towards expected performance levels. The assessment decisions that support the application of these mechanisms should guide educators on how to treat results and communicate them to students.

ICT tools are useful for conducting application and analysis of assessment results. Nowadays, educators can count on multiple computing applications and devices for planning, conducting, tracking, analyzing student's assessment results, and giving feedback. Most of these systems integrate alternatives for self and peer assessment process. Besides, new computing tools based on AI, machine learning, and electronic systems to apply crowdsource-based relevance assessment are part of these new alternatives educators can use for assessing purposes [36, 46]. The debate at this point goes around first, throughout the possibility that educators get into a misunderstanding of the role of assessment literacy while the use of electronic assessment tools become more familiar. Even if this new technology is making assessment work more efficient in data process, precision, and faster delivery of results, educators will continue to take the most appropriate decisions about assessment and learning outcomes depending on the context, modality, learning goals, and discipline.

The second point of debate refers to the use of this new generation technology in non-formal or continuous learning scenarios for assessment purposes. For instance, massive online open courses and self-learning training modules are some of the most popular self-directed learning processes for lifelong learning. Indeed, AI new technologies permit students the opportunity to adapt the learning process to their needs while taking lessons, going further, or getting back in the learning track according to their performance. Nevertheless, there is a latent risk learner could get only emphasis on summative assessment and high-stakes testing. Again, at this point, educators need to get involved in learning assessment decisions to guide the inclusion of high-quality formative assessment process for this kind of learning purpose and environment [68, 70].

1.7 Conclusions

Learning assessment is a complex task. The process of planning, design, and implementation of formative and summative assessment go further the application of activities, quizzes, and tests. These practices must be based on psycho-pedagogical fundamentals in coherence with the conjunction of personal, institutional, and external factors that permit to configure and design learning assessment environments.

In this chapter, the author gathered information from scientific studies mainly that proved, discussed, and concluded how important learning assessment conduction is and how it impacts students' academic achievement. The critical element is based on the principles of assessment literacy as the disciplinary orientation coming from psycho-pedagogical approaches that permit educators the appropriation of involved relevant: educational paradigms, educational models, own experience and beliefs, conceptual, and theoretical, pedagogical approaches and learning taxonomies. This initial information permit educators to start taking augmented decisions on how to design coherent and useful assessment environments.

Afterwards some more other decisions should also be taken along this pathway. One is related to the connection of the subject matter or discipline and the expected performance level with the assessment practices. The other is associated with the frequency, time, instruments, grading, and relevant information to be collected and treated to give results and feedback to students. One last but not less important step is the selection of ICT tools to support assessment procedures and systematization and data analysis. This selection is closely related to the needs of learners rather than just operational requirements. Undoubtedly, the pending task for the following years involves computer science professionals, and educators to work together towards ICT assessment tools to become an essential part of a learning environment, bringing practicality while conducting assessment without neglecting fundamentals and intention. This could be considered one of the most challenging tasks in the assessment literacy matter.

Finally, some discussion goes around meta-evaluation and accountability. These two practices enhance educators' assessment practice and fortify assessment environment design. Besides, permits the revision of educational goals and the improvement of instructional practices. This task also contributes to empowering credibility of institutions.

References

1. Aboalela, R., & Khan, J. (2017). *Model of learning assessment to measure student learning: Inferring of concept state of cognitive skill level in concept space* (pp. 189–195).
2. Ahankari, S., & Jadhav, A. (2018). A novel approach of software based rubrics in formative and summative assessment of affective and psycomotor domains among the engineering under graduates: Focusing on accrediation process across pan India. In *Proceedings—IEEE 18th International Conference on Advanced Learning Technologies, ICALT 2018* (pp. 426–430). IEEE.
3. Al-Azawei, A., Baiee, W. R., & Mohammed, M. A. (2019). Learners' experience towards e-assessment tools: A comparative study on virtual reality and moodle quiz. *International Journal of Emerging Technologies in Learning (iJET), 14,* 34–50. https://doi.org/10.3991/ijet.v14i05.9998.
4. Alkharusi, H. (2008). Effects of classroom assessment practices on students' achievement goals. *Educational Assessment, 13,* 243–266. https://doi.org/10.1080/10627190802602509.
5. Anderson, L. W., Krathwohl, D. R., & Bloom, B. S. (2001). *A taxonomy for learning, teaching, and assessing: A revision of Bloom's taxonomy of educational*. Logman.

6. Antoniou, P., & James, M. (2014). Exploring formative assessment in primary school classrooms: Developing a framework of actions and strategies. *Educational Assessment, Evaluation and Accountability, 26,* 153–176. https://doi.org/10.1007/s11092-013-9188-4.
7. Aydin, M., Baki, A., Köğce, D., & Yildiz, C. (2009). Mathematics teacher educators' beliefs about assessment. *Procedia—Social and Behavioral Sciences., 1,* 2126–2130.
8. Beatty, I. D., & Gerace, W. J. (2009). Technology-enhanced formative assessment: A research-based pedagogy for teaching science with classroom response technology. *Journal of Science Education and Technology, 18,* 146–162. https://doi.org/10.1007/s10956-008-9140-4.
9. Bennett, R. E. (2011). Formative assessment: A critical review. *Assessment in Education Principles Policy and Practice, 18,* 5–25. https://doi.org/10.1080/0969594X.2010.513678.
10. Black, P., & Wiliam, D. (2009). Developing the theory of formative assessment. *Educational Assessment, Evaluation and Accountability, 21,* 5–31. https://doi.org/10.1007/s11092-008-9068-5.
11. Bloom, B. S. (1956). *Taxonomy of educational objectives: The classification of educational goals.* New York: David McKay.
12. Bloom, B. S., Madaus, G., & Hastings, J. T. (1971). *Handbook on formative and summative evaluation of student learning.* New York: McGraw-Hill.
13. Brookhart, S. M. (1997). A theoretical framework for the role of classroom assessment in motivating student effort and achievement. *Applied Measurement in Education, 10,* 161–180. https://doi.org/10.1207/s15324818ame1002_4.
14. Brown, G. T. L., Harris, L. R., & Harnett, J. (2012). Teacher beliefs about feedback within an assessment for learning environment: Endorsement of improved learning over student well-being. *Teaching and Teacher Education, 28,* 968–978. https://doi.org/10.1016/j.tate.2012.05.003.
15. Castillo, M., Heredia, Y., & Gallardo, K. (2017). Collaborative work competency in online postgraduate students and its prevalence on academic achievement. *Turkish Online Journal of Distance Education, 18,* 168–179. https://doi.org/10.17718/tojde.328949.
16. Cooper, P. A. (1993). From behaviorism to cognitivism to constructivism. *Educational Technology, 33,* 12–19.
17. Dobson, J. L. (2008). The use of formative online quizzes to enhance class preparation and scores on summative exams. *American Journal of Physiology—Advances in Physiology Education, 32,* 297–302. https://doi.org/10.1152/advan.90162.2008.
18. Dunn, K. E., & Mulvenon, S. W. (2009). A critical review of research on formative assessment: The limited scientific evidence of the impact of formative assessment in education. *Practical Assessment, Research and Evaluation, 14,* 1–11.
19. Ellis, L., Marston, C., Lightfoot, J., & Sexton, J. (2015). *Faculty professional development* (pp. 69–80).
20. Febriani, I., & Irsyad Abdullah, M. (2018). A systematic review of formative assessment tools in the blended learning environment. *International Journal of Engineering and Technology, 7,* 33–39. https://doi.org/10.14419/ijet.v7i4.11.20684.
21. Fischer, E., & Hänze, M. (2019). How do university teachers' values and beliefs affect their teaching? *Educational Psychology, 40,* 1–22. https://doi.org/10.1080/01443410.2019.1675867.
22. Florian-Gaviria, B., Glahn, C., & Fabregat Gesa, R. (2013). A software suite for efficient use of the European qualifications framework in online and blended courses. *IEEE Transactions on Learning Technologies, 6,* 283–296. https://doi.org/10.1109/TLT.2013.18.
23. Giraldo, F. (2017). A diagnostic study on teachers' beliefs and practices in foreign language assessment. *Ikala, 23,* 25–44. https://doi.org/10.17533/udea.ikala.v23n01a04.
24. Glazer, N. (2014). Formative plus summative assessment in large undergraduate courses: Why both? *International Journal of Learning in Higher Education, 26,* 276–286.
25. Guskey, T. (2005). Formative classroom assessment and Benjamin S. Bloom: theory, research, and implications. In *Annual meeting of the American Educational Resaerch Association* (pp. 1–11).

26. Hancock, A. B., & Brudage, S. B. (2010). Formative feedback, rubrics, and assessment of professional competency through a speech-language pathology graduate program. *Journal of Allied Health, 39,* 110–119.
27. Harlen, W., & James, M. (1997). Assessment and learning: Differences and relationships between formative and summative assessment. *Assessment in Education Principles Policy and Practice, 4,* 365–379. https://doi.org/10.1080/0969594970040304.
28. Hattie, J., & Timperlay, H. (2007). The power of feedback. *Review of Educational Research, 44,* 16–17. https://doi.org/10.1111/j.1365-2923.2009.03542.x.
29. Hermans, R., van Braak, J., & Van Keer, H. (2008). Development of the beliefs about primary education scale: Distinguishing a developmental and transmissive dimension. *Teaching and Teacher Education, 24,* 127–139. https://doi.org/10.1016/j.tate.2006.11.007.
30. Hooshyar, D., Ahmad, R. B., Yousefi, M., Fathi, M., Horng, S. J., & Lim, H. (2016). Applying an online game-based formative assessment in a flowchart-based intelligent tutoring system for improving problem-solving skills. *Computers and Education, 94,* 18–36. https://doi.org/10.1016/j.compedu.2015.10.013.
31. Houston, D., & Thompson, J. N. (2017). Blending formative and summative assessment in a capstone subject: 'It's not your tools, it's how you use them.' *Journal of University Teaching and Learning Practice (JUTLP), 14,* 2.
32. Hutchinson, A., Moskal, B., Dann, W., & Cooper, S. (2005). Formative assessment: An illustrative example using "Alice". In *ASEE Annual Conference and Exposition, Conference Proceedings* (pp. 6521–6527).
33. Imrie, B. W. (1995). Assessment for learning: Quality and taxonomies. *Assessment and Evaluation in Higher Education, 20,* 175–189. https://doi.org/10.1080/0260293950865719.
34. James, M. (2006). Assessment, teaching and theories of learning. In J. Gardner (Ed.), *Assessment and learning* (pp. 47–60). London: SAGE.
35. Jingna, D. (2012). Application of humanism theory in the teaching approach. *Higher Education of Social Science, 3,* 32–36. https://doi.org/10.3968/j.hess.1927024020120301.1593.
36. Jordan, M. I., & Mitchell, T. M. (2015). Machine learning: Trends, perspectives, and prospects. *Science, 80*–(349), 255–260. https://doi.org/10.1126/science.aac4520.
37. Joyce, B., Calhoun, E., & Hopkins, D. (2009). *Models of learning, tools for teaching* Berkshire, England: McGraw-Hill Open University Press.
38. Kingston, N., & Brooke, N. (2011). Formative assessment: A meta-analysis and a call for research. *Educational Measurement: Issues and Practice, 30,* 28–37. https://doi.org/10.1111/j.1745-3992.2011.00220.x.
39. Knight, P. T. (2002). Summative assessment in higher education: Practices in disarray. *Studies in Higher Education, 27,* 275–286. https://doi.org/10.1080/03075070220000662.
40. Lees, R., & Anderson, D. (2015). Reflections on academics' assessment literacy. *London Review of Education (LRE), 13,* 42–48. https://doi.org/10.18546/LRE.13.3.06.
41. Leiva-Brondo, M., Cebolla-Cornejo, J., Peiró, R. M., & Pérez-de-Castro, A. M. (2017). Collaborative work and outcome assessment: A good combination. In *INTED2017 proceedings* (pp 4950–4955).
42. Madaus, G. F., & Stufflebeam, D. L. (1984). Educational evaluation and accountability: A review of quality assurance efforts. *American Behavioral Scientist, 27,* 649–672.
43. Marzano, R., & Kendall, J. (2006). *The new taxonomy of educational objectives* (2nd ed.). Thousand Oaks, CA: Corwin Press, SAGE Publication Company.
44. McDaniel, M. A., Wildman, K. M., & Anderson, J. L. (2012). Using quizzes to enhance summative-assessment performance in a web-based class: An experimental study. *Journal of Applied Research in Memory and Cognition, 1,* 18–26. https://doi.org/10.1016/j.jarmac.2011.10.001.
45. Mellati, M., & Khademi, M. (2018). Exploring teachers' assessment literacy: Impact on learners' writing achievements and implications for teacher development. *Australian Journal of Teacher Education, 43,* 1–18. https://doi.org/10.14221/ajte.2018v43n6.1.
46. Moshfeghi, Y., Huertas Rosero, A. F., & Jose, J. M. (2016). A game-theory approach for effective crowdsource-based relevance assessment. *ACM Transactions on Intelligent Systems and Technology (TIST), 7,* 1–25. https://doi.org/10.1145/2873063.

47. Nicol, D., & MacFarlane-Dick, D. (2006). Formative assessment and selfregulated learning: A model and seven principles of good feedback practice. *Studies in Higher Education, 31,* 199–218. https://doi.org/10.1080/03075070600572090.
48. Norman, W., & Steinaker, B. M. R. (1979). *Experiential taxonomy: A new approach to teaching and learning*. London: Academic Press.
49. Panadero, E., & Jonsson, A. (2013). The use of scoring rubrics for formative assessment purposes revisited: A review. *Educational Research Review, 9,* 129–144. https://doi.org/10.1016/j.edurev.2013.01.002.
50. Pastore, S., & Andrade, H. L. (2019). Teacher assessment literacy: A three-dimensional model. *Teaching and Teacher Education, 84,* 128–138. https://doi.org/10.1016/j.tate.2019.05.003.
51. Patton, M. Q. (1996). A world larger than formative and summative. *Evaluation Practice, 17,* 131–144.
52. Pombo, L., & Talaia, M. (2012). Evaluation of innovative teaching and learning strategies in science education: Collaborative work and peer assessment. *Problems of Education in the 21st Century, 43,* 86–95.
53. Pryor, J., & Torrance, H. (1997). Formative assessment in the classroom: Where psychological theory meets social practice. *Social Psychology of Education, 2,* 151–176. https://doi.org/10.1023/A:1009654524888.
54. Raupach, T., Brown, J., Anders, S., Hasenfuss, G., & Harendza, S. (2013). Summative assessments are more powerful drivers of student learning than resource intensive teaching formats. *BMC Medicine, 11,* 1–10. https://doi.org/10.1186/1741-7015-11-61.
55. Rolfe, I., & McPherson, J. (1995). Formative assessment: How am I doing? *Lancet, 345,* 837–839. https://doi.org/10.1016/S0140-6736(95)92968-1.
56. Ruochen, L. R., Kitche, H., Bert, G., Richardson, M., & Fordham, E. (2019). *OECD reviews of evaluation and assessment in education.* Georgia.
57. Rushton, A. (2009). Formative assessment: A key to deep learning? *Medical Teacher, 27,* 509–513. https://doi.org/10.1080/01421590500129159.
58. Sadid-Zadeh, R., D'Angelo, E. H., & Gambacorta, J. (2018). Comparing feedback from faculty interactions and virtual assessment software in the development of psychomotor skills in preclinical fixed prosthodontics. *Clinical and Experimental Dental Research, 4,* 189–195. https://doi.org/10.1002/cre2.129.
59. Sadler, D. R. (1989). Formative assessment and the design of instructional systems. *Instructional Science, 144,* 119–144.
60. Sadler, D. R. (1998). Formative assessment: Revisiting the territory. *Assessment in Education Principles Policy and Practice, 5,* 77–84. https://doi.org/10.1080/0969595980050104.
61. Saeed, M., Tahir, H., & Latif, I. (2018). Teachers' perceptions about the use of classroom assessment techniques in elementary and secondary schools. *Bulletin of Educational Research, 40,* 115–130.
62. Sanusi, N. M., Kamalrudin, M., & Mohtar, S. (2019). Student engagement using learning management system in computer science education. *International Journal of Recent Technology and Engineering, 8,* 743–747. https://doi.org/10.35940/ijrte.B1121.0982S1119.
63. Schmid, R. (2018). Pockets of excellence: Teacher beliefs and behaviors that lead to high student achievement at low achieving schools. *SAGE Open, 8,* 215824401879723. https://doi.org/10.1177/215824401879723s8.
64. Scopus. (2020). Bibliometric data from Scopus database.
65. Scriven, M. (1967). The methodology of evaluation. In R. W. Tyler, R. M. Gagne, & M. Scriven (Eds.), *Perspectives of curriculum evaluation* (pp. 39–83). Chicago: Rand McNally.
66. Shepard, L. A. (2000). The role of assessment in a learning culture. *Educational Researcher, 29,* 4–14. https://doi.org/10.3102/0013189X029007004.
67. Simper, N. (2018). Rubric authoring tool supporting cognitive skills assessment across an institution. *Teaching and Learning Inquiry, 6,* 10–24. https://doi.org/10.20343/teachlearninqu.6.1.3.
68. Spector, J. M., Ifenthaler, D., Sampson, D., Yang, L. J., Warusavitarana, A., Dona, K. L., et al. (2016). International forum of educational technology & society technology enhanced

formative assessment for 21st century learning linked references are available on JSTOR for this article: technology enhanced formative assessment for 21st century learning. In *International Forum of Educational Technology & Society is Collaborating with JSTOR to Digitize, Preserve and Extend Access to Journal of Educational Technology & Society*, 57–71.
69. Sridharan, B., Tai, J., & Boud, D. (2019). Does the use of summative peer assessment in collaborative group work inhibit good judgement? *Higher Education, 77,* 853–870. https://doi.org/10.1007/s10734-018-0305-7.
70. Steffens, K., Bannan, B., Dalgarno, B., Bartolomé, A. R., Esteve-González, V., & Cela-Ranilla, J. M. (2015). Recent developments in technology-enhanced learning: A critical assessment. *RUSC. Universities and Knowledge Society Journal, 12,* 73. https://doi.org/10.7238/rusc v12i2.2453.
71. Stufflebeam, D. (2011). Meta-evaluation. *Journal of MultiDisciplinary Evaluation, 7,* 99–158.
72. Stufflebeam, D. L. (2000). The methodology of metaevaluation as reflected in metaevaluations by Western Michigan University Evaluation Center. *Journal of Personnel Evaluation in Education, 14,* 95–125. https://doi.org/10.1023/A:1008198315521.
73. Taras, M. (2005). Assessment—Summative and formative—Some theoretical reflections. *British Journal of Educational Studies, 53,* 466–478. https://doi.org/10.1111/j.1467-8527.2005.00307.x.
74. Taras, M. (2008). Summative and formative assessment: Perceptions and realities. *Active Learning in Higher Education, 9,* 172–192. https://doi.org/10.1177/1469787408091655.
75. Taras, M., & Davies, M. S. (2017). Assessment beliefs of higher education staff developers. *London Review of Education (LRE), 15,* 126–140. https://doi.org/10.18546/LRE.15.1.11.
76. Tedesco, J. C. (2016). *Ten notes on learning assessment systems.*
77. Tsai, F. H., Tsai, C. C., & Lin, K. Y. (2015). The evaluation of different gaming modes and feedback types on game-based formative assessment in an online learning environment. *Computers and Education, 81,* 259–269. https://doi.org/10.1016/j.compedu.2014.10.013.
78. Tünnermann, C. (2008). *Modelos educativos y académicos.* Managua, Nicaragua: Editorial Hispamer.
79. Wang, T. H. (2008). Web-based quiz-game-like formative assessment: Development and evaluation. *Computers and Education, 51,* 1247–1263.
80. Widiastuti, I. A. M. S., Mukminatien, N., Prayogo, J. A., & Irawati, E. (2020). Dissonances between teachers' beliefs and practices of formative assessment in EFL classes. *International Journal of Instruction, 13,* 71–84. https://doi.org/10.29333/iji.2020.1315a.
81. Hitchel, H. J., Claxton, H. L., Holmes, D. C., Ranji, T. T., Chalkley, J. D., Santos, C. F., et al. (2018). A trigger-substrate model for smiling during an automated formative quiz: Engagement is the substrate, not frustration. In *ACM's International Conference Proceedings Series* (*ICPS*). https://doi.org/10.1145/3232078.3232084.
82. World Bank. (2019). *Classroom assessment: Taking the first steps towards improved teaching and learning in Tajikistan*
83. Xiang, J., & Ye, L. (2009). A general software framework based on reform in formative assessment. *Journal of Software, 4,* 1076–1083. https://doi.org/10.4304/jsw.4.10.1076-1083.
84. Yorke, M. (2003). Formative assessment in higher education: Moves towards theory and the enhancement of pedagogic practice. *Higher Education, 45*(4), 477–501. https://doi.org/10.1023/a:1023967026413.
85. Zainuddin, Z., Shujahat, M., Haruna, H., & Chu, S. K. W. (2020). The role of gamified e-quizzes on student learning and engagement: An interactive gamification solution for a formative assessment system. *Computers and Education, 145,* 103729. https://doi.org/10.1016/j.compedu.2019.103729.

Chapter 2
Exploring Collaboration and Assessment of Digital Remote Teams in Online Training Environments

S. Chujfi, H. Traifeh, T. Staubitz, R. Refaie, and C. Meinel

Abstract Our digital and geographically distributed society faces great challenges nowadays in managing communication between people—who in most cases—have never met each other before. This is particularly relevant for institutions working with peers that interact in remote environments where team collaboration is an essential driver of strategic innovations, and which stresses the need to explore different techniques to enable teamwork and synergy between those—unknown—digital peers. Training students and employees to be able to manage the challenges that result from this kind of work become increasingly important. Therefore, the development of online assessments is mandatory to encourage virtual participation and teamwork analysis in order to boost objectivity and mark the outcomes of teamwork. Following a summary of the most challenging aspects that remote teams face, this chapter describes how the self, and peer-assessment strategies defined by Willcoxson can be successfully replicated in geographically decentralized teams using online tools. They can be used to develop team dynamics to accurately assess individuals' contributions to teams, and also project planning and resourcing. The concept is determined by the transparency of team profiles and the definition of roles within the remote teams to encourage high responsiveness of all members. Conclusively, the evaluation approach should be targeted to be used to improve effective remote

S. Chujfi (✉) · H. Traifeh · T. Staubitz · R. Refaie · C. Meinel
Hasso Plattner Institute for Digital Engineering, University of Potsdam, Potsdam, Germany
e-mail: Salim.Chujfi@hpi.de

H. Traifeh
e-mail: Hanadi.Traifeh@hpi.de

T. Staubitz
e-mail: Thomas.Staubitz@hpi.de

R. Refaie
e-mail: Reem.AbouRefaie@hpi.de

C. Meinel
e-mail: Christoph.Meinel@hpi.de

R. Babo et al. (eds.), *Workgroups eAssessment: Planning, Implementing and Analysing Frameworks*, Intelligent Systems Reference Library 199,
https://doi.org/10.1007/978-981-15-9908-8_2

teamwork and also develop planning abilities, as well to measure whether the collaboration and communication executed virtually were truly successful. We have applied and evaluated our approach during a Massive Open Online Course where the participants learned about effective virtual teamwork strategies. About 3000 participants were enrolled in the course, about 50% of them have been active in the course. The course ran over 4 weeks and included a hands-on task that had to be solved in a virtual team.

Keywords Online assessments · Digital remote teams · Online collaboration · Massive open online course · Virtual teamwork strategies

2.1 Introduction

Organizations are continuously working with digital remote teams to be more active and flexible as part of their tactical modernization, with numerous benefits to society, institutions, and people. Reports from English speaking countries, Eastern Europe and China emphasize that the ability to work in teams and the "related interpersonal skills are equally or more important than graduates' technical skills" [1]. This is also reflected in an increasing demand for digital teamwork skills on the job market as shown by a poll by the Association of American Colleges and Universities (AACU) in 2009, whereby 71% of the employers urged colleges to place greater emphasis on teamwork skills [2].

Nevertheless, such transformations may affect the effective knowledge exchange, particularly, because in geographically dispersed sceneries and in low-collaborative environments, individuals are not able to use gesticulations or natural body language expressions and have to alter their languages without obtaining any response, which may disturb the foundation for creating knowledge.

Teamwork has a different dimension in remote environments and needs effective conventions to combine information, that is coming from the individual his social context and also from the content itself. Moreover, there are social dynamics that materialize while interacting online and are also relevant to consider. Several academics acknowledged already the challenges that remote environments and teammates collaboration may induce [3]. For example, the impact of confidence and strength required of developing it in remote settings has been exhaustively considered [4]. Consequently, some academics have suggested different methods, such as those which examine the functioning network definition of decentralized teamwork [5]. Others explore the relations and interaction among individuals within remote teams [6].

Since different remote team members with diverse skills and knowledge belong to remote teams, it is highly relevant to consider matching the backgrounds of the participants. Previous knowledge, working experiences, geographic locations, availability, time commitments, and also cognitive and behavioral tools allow us to define conducting strategies of team members and set behavioral guides of interaction, allowing improved communication and effectiveness.

Remote learning conditions are significant to corroborating the issues in a systematic way from numerous viewpoints (specialties). This may drive analysis to an examination, improving experimentations, and situations and is consequently contemplated as the focus of our methodology. Getting in continuous discussions between persons with diverse backgrounds is the essence and immense chance of enhancing distance teammates' knowledge.

This work analyzes collaboration in online training environments with decentralized individuals, assigning individual and collaborative activities with the flexibility to use different tools to interact. It also evaluates grouping strategies to analyze online behaviors of team members within their online environments, so as to effectively classify personalities in order to evaluate greater efficacy of communication.

To do so, it draws on evidence from the course titled "Introduction to Successful Remote Teamwork", offered on openHPI, which is an educational Internet platform of the German Hasso Plattner Institute in Potsdam. The course covers the benefits and risks of driving a virtual team culture and how guided remote work drives international teams to success. The course also involves the use of intercultural competence as a key factor of interaction and communication. In the hands-on part of the course, the participants learn how to select appropriate online collaboration tools and how to employ them in a practical task. Working with a "real life" virtual team, the participants gain first-hand experience about the opportunities and challenges of tele-working. To assess the group dynamics and provide insights on how to optimize them, the course applies Willcoxson's methodology, which is a comprehensive assessment strategy that translates individual team input into points awarded, and at the same time, allows for regular oversight on asserted influences to the product and development of collaboration.

2.1.1 Context: Relevance and Challenges of Remote Teams

A main benefit of remote teams is how they profit out of the knowledge of dispersed individuals with diverse qualifications and expertise [7, 8]. Nevertheless, the foundation of remote societies using computer-mediated communication (CMC), has not always received positive appraisals due to its narrow ability to transfer rich information and its subsequent unsuccessfulness in transmitting non-codified knowledge [9].

The implication of managing knowledge within the distributed or discontinuous remote team environments arises immense challenges on today's institutions that are focused to obtain greatest effectiveness. Table 2.1. Presents some very challenging factors that remote team members face today when performing tasks remotely. Irrespectively of the technical equipment available, many different aspects seem to have a close relation to cognitive and rational preferences that are linked to the individuals and which are not examined, evaluated, or even acknowledged; certainly, even

Table 2.1 Challenges facing staff when working remotely [10]

Challenges	Description
Lack of self-discipline	Staff not confident to perform assigned work effectively
Lack of self-management	Staff not able to manage themselves
Lack of organizational engagement	Staff do not feel a part of the organization
Lack of coordination	Staff lack direction and instructions from management
Lack of motivation	Staff lack empowerment and feedback
Lack of feedback	Staff lack information to continue performing activities
Social isolation	Staff lack contact with others
Distractions	Staff not able to focus on assigned work
Slow communication	Staff not able to effectively communicate
Work-life balance	Staff not able to separate private and work life
Ambiguity	Staff confused about how to proceed due to vagueness
Less structure	Staff lack organizational structure

being aware that personal and direct communication does not occur as it happens in traditional situations.

Considering the complexity that remote work encompasses, not only the roles of each team member are important, but also the topologies of interaction are imperative while managing geographically distributed members. Functional roles are relevant for enabling the team to achieve its project objectives. People with expertise and management skills are the most common profiles expressing this role. Socio-emotional roles help to build trust within the team, as well as support teams to collaborate and effectively work to achieve common goals. Individual roles are relevant for reflecting the personal needs of an individual such as desire for recognition or control. These are considered to have adverse effects on trust and collaboration if are not properly managed.

Figure 2.1 presents three different topologies on how interaction may happen remotely, and also require considering the need for the previously mentioned roles so that interaction happens, trust is built, and the goals of the teamwork are achieved. The selection of the appropriate topology of interaction in a team is mainly decided based on the specialty of the work as well as the size of the team.

Fig. 2.1 Topologies of remote interaction

Jehng [11], Civin [12], Fusaro [13], and Wellman [14] indicate that the unclear, self-contained, and less visibly systematized settings where digital media are implemented with the purpose of teamwork, undesirably affect remote team worker's commitments regarding their professions and their geographically distributed teammates. It is important to mention little remote participation and engagement is not pervasive among individuals [15]. Cerulo [16] and Ngwenyama [17] emphasize that a digital dominated atmosphere may develop mindsets of self-determination for greatest performance, as well as remote team mates could adopt spontaneous affection and camaraderie with remote coworkers [16, 17].

Several factors are known to affect predilections for learning unaccompanied versus learning in groups, those particularly related to social individuality and collectivism [18], the discrepancy of awareness between distance work and remote colleagues is partially related to the cognitive coherence between people's and the clearly perceptible characteristics of the digital resources available for collaboration with their remote colleagues in order to accomplish their tasks [19, 20].

Additionally, the activity of gathering and handling knowledge in geographically distributed spaces is accomplished with the assistance of digital resources that empower individuals to communicate and share information. Considering a richer medium represents also an improvement for better collaboration which consequently can enrich individuals comprehensibility among collaborative participants [11, 21]. The digital vividness and depth giving to content, assist to decline the sense of isolation and the insecurity created by digital mediation environment and interactive circumstances. This alters the way individuals approach their responsibilities and drive them to compare them to each other and eventually influences their motivation and the motivation of the remote teammates [22, 23].

Scientists have assessed numerous approaches to evaluate sentiments and emotions, to identify temperaments and reactions, and conclusively, to recognize the cause, intention, and multifaceted assertiveness types [24]. Different approaches have tried to study influences, trying to recognize the latent effects of group collaboration developments on each other and on participants. Some contemporary practices have developed more challenging approaches, such as crowding, to openly measure the possible influence between variables [24].

The relevance of distinguishing conducts and collaboration shapes within teammates could be associated to hypothetical collective intelligence coherence grounded

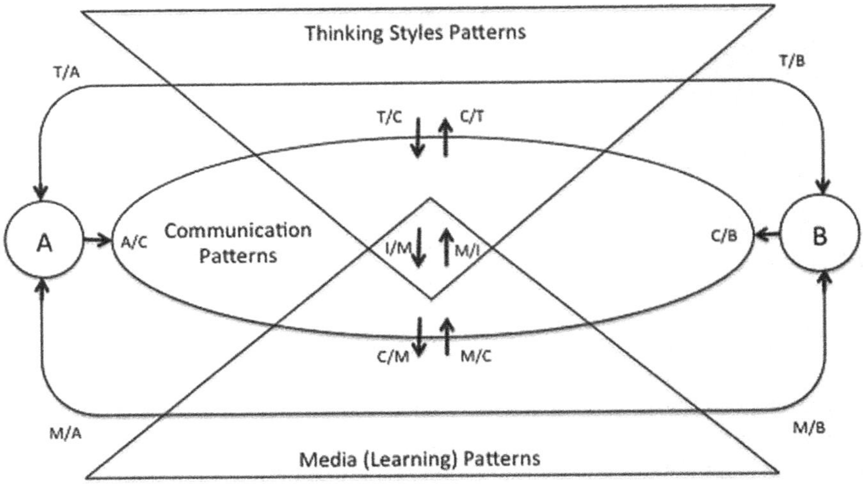

Fig. 2.2 Three different stages of interaction [26]

on cognitive predilections. Contemplating the McGrath [25] interaction practices at a detailed level, Fig. 2.2 illustrates the collaboration in three diverse phases: communication shape amongst interacting individuals, subject of communications, together with effect of diverse group collaboration methods on each other and on individuals.

A/B:	Group Members.
Form:	A/C, C/B: Online Communication process.
Content:	T/A, T/B: Interpersonal activity (Thinking).
	M/A, M/B: Media activity (Learning).
Outcome:	T/C, C/T: Effect of Communication pattern and interpersonal component.
	C/M, M/C: Effect of Communication pattern and task component.
	I/M, M/I: Effect of interpersonal and task component on one another.

2.1.2 *Willcoxson Methodology*

Willcoxson's [27] methodology responds to the need for a comprehensive assessment strategy that translates student input into marks awarded and at the same time allows for staff checks on claimed contributions to the product and process of teamwork. It holds that regardless of the field of work, a comprehensive assessment strategy requires peer and self-assessment not just of team dynamics, but also of project management activities.

Hence, the proposed tool attempts to develop and assess project management and resourcing strategies, concurrently with team dynamics, instead of sequentially. This serves as a basis for a subjective assessment (of team dynamics and relative input

Table 2.2 Willcoxson's methodology

Phases	Main outcomes
1	Engage in individual reflection of the negative teamwork experiences
2	Agree on a set of rules to govern team interactions
3	Produce a comprehensive project plan
4	Reflect on the relationship between what was planned and what happened in the area of project management

into the team report) but also for objective assessment related to tasks contracted and tasks completed.

To accomplish this, the methodology is divided into four distinct phases, which provide a standardized, simple structure for formative development of task and interpersonal behaviors, for the cumulative assessment of these, and for the collation of relevant information needed to discover discrepancies between self and peer reporting of behaviors. Each phase of the tool is presented in an easily digestible one sheet of paper (or more, if needed to capture relevant information), and handed to the students at the appropriate time. When presenting the tool, comments are provided to assist users in effectively utilizing it (Table 2.2).

Phase one consists of individual reflections on the negative teamwork experiences, which aims at specifying and/or documenting the main elements that make a poorly performing groups or teams. The main premise behind this phase is that the request for information about bad experiences arises from the personal observations that students have ready access to memory associated with strong negative emotions. This makes it easy for them to narrate the internal and external causes of these negative emotions, and thus provide a clear perspective of their own and others' responsibilities in group situations. Willcoxson argued that asking about good teamwork experiences makes it more challenging for students to identify specific issues and engage critically with their causes. This often leads to blame-shifting and attributing the responsibility for group development and/or maintenance on others.

Phase two sets the 'group work' rules, whereby the students engage in comparing the bad experiences they had individually iterated in phase one, with the elements that they envision will make a group really good to work with. The purpose of this phase is to openly discuss as a team and compare between the elements that make a group good and the elements that make it bad; and to ultimately reach an agreement on the rules that govern their team interactions.

Phase three comprises of the group project planning sheet, which outlines three main segments designed to enable them to scope the project, distribute tasks, and determine the resources needed for implementation.

1. All the stages involved in completing the project, the time needed to complete each stage, and the group members who will be responsible for the implementation of assigned tasks.

2. Additional information, resources, or help needed to complete the project, where or how to find resources, and which group members will be tasked with acquiring them.
3. Ways to ensure that all team members have an overview of the whole project and understand the underlying principles.

Phase four outlines the group attachment chart, which guides students into a series of reflections on the relationship between what was planned and what happened in the area of project management with a special focus on the extent to which team members fulfilled their obligations, and what were the lessons learned from undertaking the team project.

2.2 Research Methodology

This chapter attempts to analyze online team collaboration with the team members being decentralized individuals, who assign individual and collaborative activities with the flexibility to use different tools to interact and communicate virtually. It also evaluates grouping strategies to investigate online conduct of individuals and their communities, so that it is possible to efficaciously recognize factors that assist the evaluation of greater success in digital communication. In doing so, the "Introduction to Successful Remote Teamwork" course was intentionally designed for the purpose of testing the Willcoxson approach to show the extent of its adaptability to "remote teams" and establishing a better understanding of how to evaluate communication both within teams and across different teams. In this section, we present the course and the design of the team assignment.

2.2.1 Course Description

Tele-working is becoming more and more a popular topic amongst modern organizations. However, it also comes with some challenges for both: tele-workers and management. The course titled "Introduction to Successful Remote Teamwork" addressed those challenges by ensuring that the participants were fit for virtual collaboration in geographically distributed contexts. It covered the benefits and risks of driving a virtual team culture and how guided remote work drives international teams to success. Furthermore, the course involved the use of intercultural competences as a key factor of interaction and communication. In the hands-on part of the course, participants learned how to select appropriate online collaboration tools and how to employ them in a practical task. Working with a "real life" virtual team, the participants gained first-hand experience about the opportunities and challenges of tele-working.

The course targeted three main groups: (1) Lower, Middle, and Senior Management Professionals who work remotely, (2) Researchers working in international projects, and (3) Everybody who is interested in collaborating with remote partners. The language of instruction was English, and the course rolled out over the duration of four (4) weeks with three (3) to six (6) active hours per week. About 3000 participants were enrolled in the course and about 50% of them have been active in the course.

The participants were given the choice between conducting a team assignment or sitting for a final multiple-choice exam, which amounts to 55% of the overall course grade. However, they were encouraged to complete both assignments to acquire the course certification.

The first week was intentionally designed to prepare the participants for the course. It covered a discussion of the meaning of remote work and contexts in which it is appropriate or not. This was followed by an in-depth reflection on the reliability of remote work for organizations, including infrastructure and data protection issues, which are elements that often discourage people from trusting online communication tools. Furthermore, it outlined a set of tools to prepare the participants for the practical tasks in the course, as well as their daily 'remote working' life.

The second week focused on "The human side of remote work" and it particularly considered two aspects that help define how the management of the remote environment can be addressed. This entailed the consideration of the characteristics of personal profiles to build and empower remote teams. And followed by defining some guidelines to understand the dimension of the remote interaction, and how can organizations successfully overcome the challenges of the remote environment.

The third week focused on the meaning of "intercultural competence" and why it is important to have such a competence in our diverse environments today; especially when working in a "remote team". It also introduced the different elements of intercultural competence and guided the participants through some practical exercises that were designed to help them better understand, as well as critically engage with their own culture, and also other cultures.

The fourth and final week was dedicated to hands-on work on the final team project. The team project was designed in the form of a challenge to provide a fictional business owner named "Alex", with a strategy of how to optimize his office space in order to ensure that his team is working comfortably and productively. This was conducted in parallel to the ongoing discussions amongst the participants that took place on a virtual collaboration space inside the platform (called *Collab Space*), which was especially designed to endorse transparent, optimize open communication among the participants and increase forum activity [2]. The Collab Space comes with a set of built-in comprehensive features that include file sharing, a team forum, a video chat App, an Etherpad, and an integrated Peer Assessment Tool, which proved to be the most successful approach according to previous studies comparing the usage of a built-in platform forum versus other social media tools [28]. The collation of all these features in a one-stop-shop was intentionally designed to save time and ensure the flow of communication within teams and across different teams [28] (Table 2.3).

Table 2.3 Course content

Course duration (weeks)	Content
Week 1	• The meaning of remote work and contexts in which it is appropriate or not • The reliability of remote work for organizations, including infrastructure and data protection issues • The relevant tools to aid participants in the practical tasks
Week 2	• The human side of remote work • Aspects that define how the management of the remote environment can be addressed
Week 3	• The meaning of "intercultural competence" and why it is important for remote teams • The different elements of intercultural competence and some exercises that help participants better understand their own culture, as well as learning how to connect with people from other cultures
Week 4	• Hands-on work on the final team project

2.2.2 Adapting Willcoxson Approach in the Team Assignment

The methodology applied in the course is an adaptation of the Willcoxson approach from the educational/classroom context to the "remote teams" context. As illustrated in Table 2.2, Willcoxson's methodology for a team process assessment tool was rolled out in four phases. However, for the use in remote teams, the course showcases an assessment strategy that is implemented in five phases. The addition of the fifth phase was necessary to allow students to re-team after the first task. This was intentionally designed in the course assessment strategy in order to make up for the fact that the participants had never met before and so it was necessary to allow some room for flexibility, which was not required for Willcoxson's setting. Another tweak to the Willcoxson method is a warm-up game that was designed for the purpose of aiding participants in getting to know each other better. Participants were asked to contact their teammates and fill a team profile matrix detailing the team members' pseudonyms, age, gender, expertise and supporting skills, meeting time preferences, leadership preferences, technological resources available, and previous experience. This is a steppingstone to generating a full-fledged team profile as it allows participants to have a better idea about each other's interests and preferences before agreeing on the ground rules.

In other respects, the methodology applied in the course bears similarities to Willcoxson's approach as it follows the same sequence and accomplishes the same outcomes. Below is a description of the Willcoxson approach as captured by "Introduction to Successful Remote Teamwork" Course. Each phase is divided into several tasks, which were either implemented individually or in groups. To ensure smooth implementation, the platform provided an annotation to guide participants through the execution of each task.

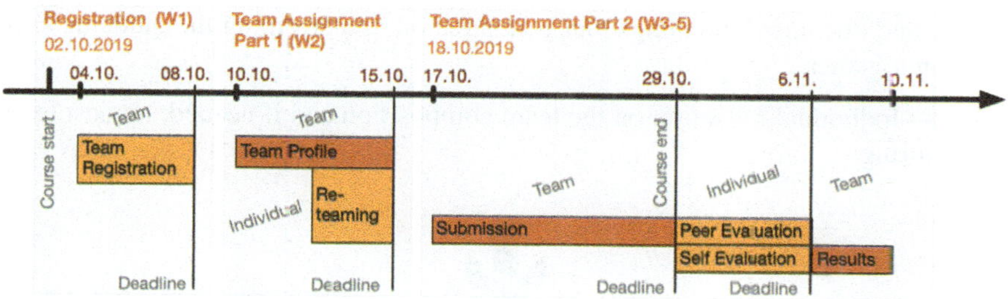

Fig. 2.3 Team assignment timeline

Figure 2.3 shows the timeline for the central elements of the team assignment. This visualization was also shown to the course participants. During the course, the image was regularly updated to illustrate which steps had already passed.

Phase 1: Individual Reflection and Team Registration

Task 1—Individual reflection on the negative experiences of working in a group or team.

Task 2—Team registration according to the participants' age, gender, time zone, area of expertise, time commitment, and leadership preference (to lead or to follow).

Corresponds with the Willcoxson's first phase, which aims at guiding the particpants into an individual reflection of their personal experiences on teamwork.

Phase 2: Team Profile

Contact the teammates and get to know each other better and fill the team profile matrix detailing the team members' pseudonyms, age, gender, expertise and supporting skills, meeting time preferences, leadership preferences, technological resources available, and previous experience.

Functions as a preparatory step for Willcoxson's second phase, which aims at reaching an agreement on the rules to govern the teams' interactions.

Phase 3: Virtualize and Verify Team Structure

Task 1—Compare the points that you have written about poorly performing teams and discuss the elements that would make a team really good to work with and be a part of, and document the things that you agree on, thus forming the guidelines for working together.

Task 2—Individual reflection on the team composition and if needed, request team restructuring.

Corresponds with the Willcoxson's second phase, which aims at reaching an agreement on the rules to govern the teams' interactions.

Phase 4: Team Project Implementation Plan

Task 1: Define all the stages involved in completing the team project, when you expect to complete each stage, and which group members will be involved in which stages (**Stage, Complete by (Date), Person(s) Responsible**).

Task 2: What additional information, help, or resources you need to complete the project. Where, or how you will find them. And which team members will be responsible for finding them.

Task 3: What will you do to make sure that every member of your group has an overview of the whole project and understands the underlying principles?

Corresponds with the Willcoxson's second phase, which aims at reaching a comprehensive project planning document to guide the team in the implementation of the team assignment.

Phase 5: Tackle the "Help Alex" Challenge

Task 1—Reflect together on the decisions taken during the planning process and how to go about completing the team project.

Task 2—Document the contributions made by the team members corresponding to the project implementation plan, while noting any differences between the planned responsibilities and what actually happened, and why these occurred. (while referencing the stages in the project, and the additional information and/or resources needed).

Task 3—Ensure that all team members have an **overview** of the project and understood the underlying principles.

Task 4—Comment on the extent to which your team kept to its guidelines for working together, which guidelines—if any—you found difficult to keep to, and why. What are the most important things you have learned from doing this project? Having completed this project, what questions remain unanswered for you?

Corresponds to Willcoxson's fourth phase, which guides the participants into reflecting on the overall performance of their teams both at the individual level and at the group level, while highlighting lessons learned.

2.2.3 *Team Building Criteria and an Overview of Participants' Profile*

The course instructors applied several team building criteria to ensure the effectiveness and harmony among teams. Contemplating that diverse team members with distinctive skills and expertise belong to remote teams, it is highly relevant to consider matching the backgrounds of the participants. The criteria work on matching participants based on previous knowledge, working experiences, geographic locations, availability, time commitments, and also cognitive and behavioral tools. This was especially designed to allow "remote teams" to define conducting profiling of teammates and propose rules of collaboration, allowing improved communication and effectiveness.

First, the participants were categorized based on their resident time zone, which was needed to match the participants with others from their time zone to make sure that meetings will be convenient for all. Nevertheless, this criterion turned to be less relevant as almost all the participants were from Germany, with a limited number of participants residing in Australia, Japan, South America, and India. This lack of cultural diversity limited the knowledge generated from culturally heterogeneous teams. Yet, this homogeneity in time zones was still useful in easing the organization of meetings and hence boosted the frequency of communication and feedback loops within teams.

The second and -arguably to be- the most important criterion for team building was the time commitment or the hours that the participants are willing to dedicate to the course. This was meant to ensure that participants would be equally committed to the task at hand, and therefore, increase the harmony among the team members. Figure 2.4 compared between time commitment distribution in the targeted course in 2019, and several MOOCs that have been conducted previously between 2016 and 2018. This suggests that building homogeneous teams was more possible in 2019 as a result of the participants being evenly distributed between 1–2 h per week and 3–4 h per week. Only a minority were willing to commit 5–6 h per week. Whereas experience from 2018 suggests that it was much more difficult to form consistent groups.

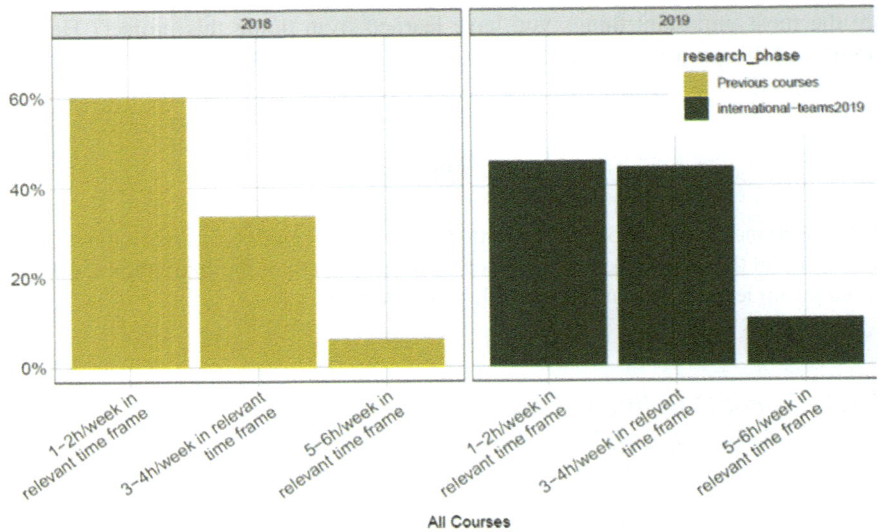

Fig. 2.4 Time commitment distribution

The third criterion looked at the leadership preference among participants, which was either the preference to lead or to follow. This criterion was applied to make sure that each team had a leader necessary for accountability and fulfilment of the tasks assigned to each team. It was also imperative to ensure that participants are comfortable in the roles assigned to them.

Gender balance was another important criterion to ensure the diversity of teams. Fulfilling this criterion was challenging as 65% of the participants were male, while only 35% were female. To overcome this hurdle, the instructors fulfilled gender balance by making sure that each team had to have at least two female members. This was derived from existing literature [29] which suggests that singular female team members tend to be inactive in team interactions, and thus, do not guarantee the integration of female input in the final team assignments.

Other diversity criterion was the variety in professional background, which turned out to be difficult to achieve as the great majority of the participants were professionals and only a minority were students, teachers, academics, or belonged to other backgrounds. Figures 2.5 and 2.6 reflect the distribution of homogeneous teams of senior professionals with over 10 years of work experience amounting to more than two-thirds of the participants (69%). This points to a higher chance that the participants will apply the knowledge gained from the course in their day-to-day work. It can also be perceived as an indication of a pressing need for such "remote teamwork" courses among professionals, hence the high enrollment statistics.

Furthermore, taking a closer look at the professionals' area of work, Fig. 2.7 shows that almost half of the participants (48%) work in IT, which means that for half of the participants, they could find others from their professional backgrounds.

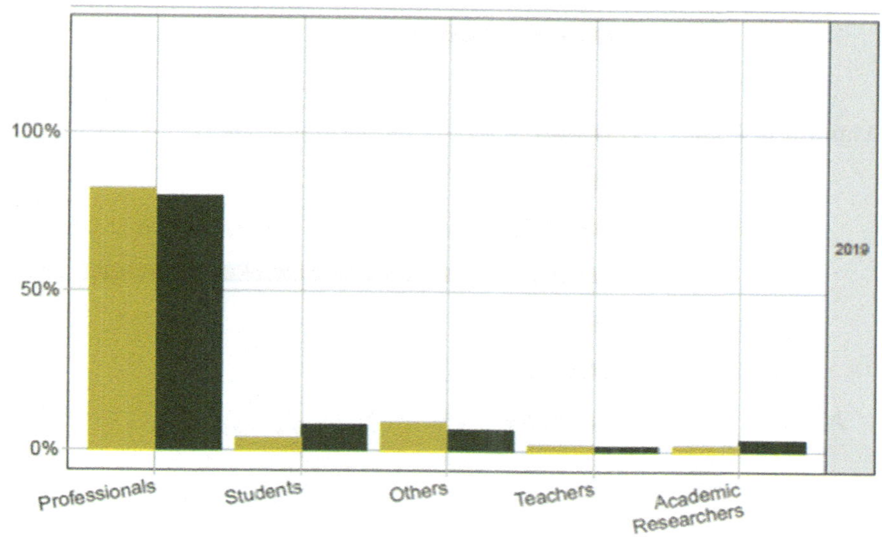

Fig. 2.5 Professional background of participants

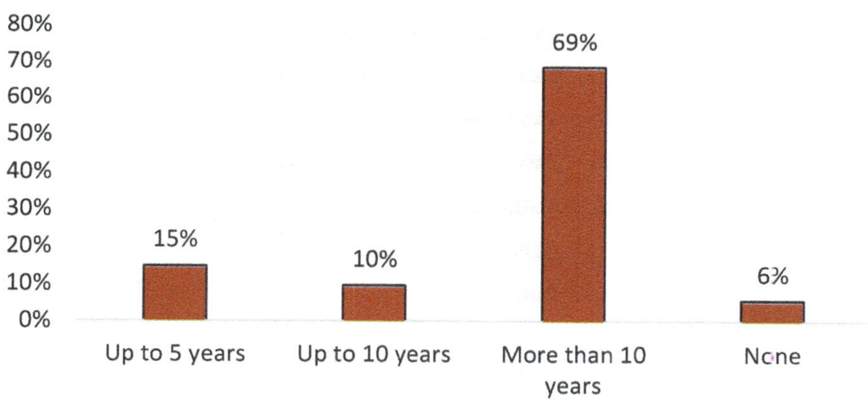

Fig. 2.6 Professional experience in team composition

Other big areas of work represented are engineering (17%), Education (11%), and Business Administration (11%). To a lesser extent marketing/sales, research, human resources, and health sectors were also represented. Of those who are employed, 18% are lower management, 17% are mid-level management, 15% identified themselves as technicians and 11% are analysts. Figure 2.8 highlights other leadership positions ranging from administrative roles (9%) and top management (7%) to supervisors (5%) and trainees (3%).

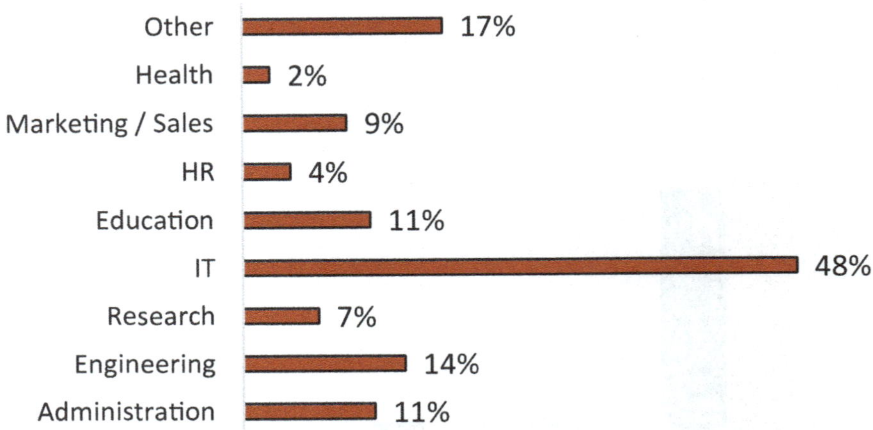

Fig. 2.7 Participants area of work

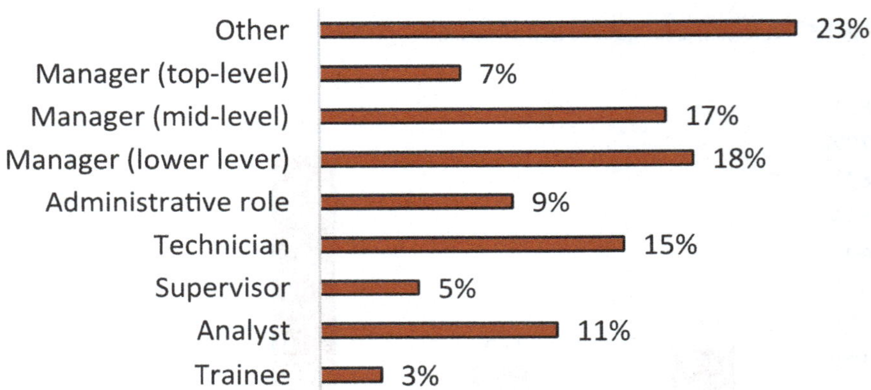

Fig. 2.8 Current employment position

Lastly, the participants were divided according to their age groups. This was successfully applied to form over 70% heterogeneous even teams that represent all age groups as illustrated by Fig. 2.9.

2.2.4 *Teams Building Method*

Those criteria were entered through a team-building software that has been developed by the openHPI team. This software matched a total of 68 teams, each comprised of 6 members, with a few teams falling below this threshold. The instructors made

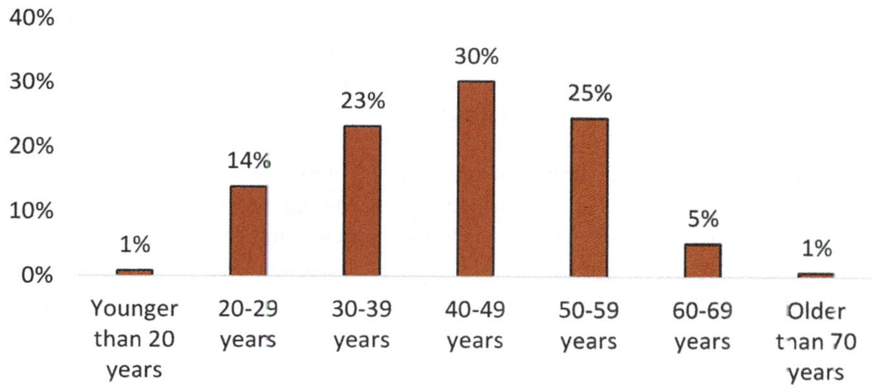

Fig. 2.9 Distribution of age in team composition

some final manual adjustments to the automatically matched teams to fine-tune the automated matching. In terms of building the teams, the course followed a strictly interventionist approach, meaning that the instructors took care of matching the team members as opposed to a laisser-faire approach, meaning the participants are selecting their teammates. Surveys in previous courses have shown that this method is preferred by the vast majority of participants [30].

2.3 Evaluation

2.3.1 Evaluation Methodology

The course combined a hoard of approaches to assess how the self, and peer-assessment strategies defined by Willcoxson methodology were replicated in geographically decentralized teams using online tools. In addition to evaluating the extent to which the assessment strategy was able to develop effective remote teamwork, was also mandatory to evaluate if the collaboration and preparation that took place virtually was really successful. To achieve this, the course followed a pre-post evaluation design, whereby the participants responded to two sets of online survey questions. The pre-survey was designed to better understand the participants' profile and overall motivation for learning about virtual teamwork before taking the course. The post-survey aimed at gauging insights on the participants' experience in the course, the level of knowledge gained, and the extent to which they expect to use this knowledge after the end of the course.

This was combined with a systematic analysis of the user interaction data throughout the duration of the course. By the end of the course, the participants also were asked to participate in a reflective exercise of "I like..." and "I wish...", that was designed to provide them with an open forum to share their perspectives of the course beyond the specific elements dictated by the post-survey.

Moreover, drawing on Bloom's grading taxonomy, the participants were given the opportunity to peer-review each other's work. This was rolled out in the form of 'Peer Assessment Workflows', which feature rubrics, where the participants are able to assign points to the submitted work corresponding to their perceived quality of the output. Each individual was tasked with producing around 3–5 reviews, thus in turn, providing each user with 3–5 reviews, which count toward the final grade. And to ensure fair evaluation, the reviewed teams had the liberty to rate the reviews they received from their peers.

2.3.2 *Outcome (Data Analysis) of the Pre-survey*

Participants' Experience with Remote Telework Before the Course

The pre-survey showed that 69% of the participants in the course have a traditional office setting where they have regular face-to-face contact with colleagues. This indicates that the majority of the participants did not have limited to no experience with telework before joining the course, and therefore, are less likely to have clear expectations. Nevertheless, a group representing 32% of the participants have worked in telework with one or more colleagues. 27% have never worked remotely before, which again emphasizes the limited prior knowledge among participants. It is also worth mentioning that 17% are used to teleworking alone. This was also reflected in their responses to enumerate the number of times they worked in virtual teams, which was as follows: 33% tried it 1–2 times; 21% of the participants work virtually all the time, and 19% attested to working virtually 3–5 times before.

Although most of the participants did not have deep experience in telework, 67% of them stated that they think it is an effective way to work. This implies that regardless of personal experience, most participants had a positive outlook on teleworking before they took the course. However, when asked whether telework should be permanent, 68% of the participants said 'no'; it should only be applied on a few days per week. More optimistic views represented 18% of the participants responded positively affirming the possibility of partaking in remote work on a full-time basis with regular virtual meetings. Unsurprisingly, 14% of the participants said that they do not know if it should be permanent or not, which could be a result of their lack of experience as Fig. 2.10.

2.3.3 *Outcome (Data Analysis) of the Post Survey*

General Course Rating and Reflections

The results of the post-survey were indicative of positive "remote team" experiences. Around 60% of the participants rated the course materials at 4 out of 5, which points to

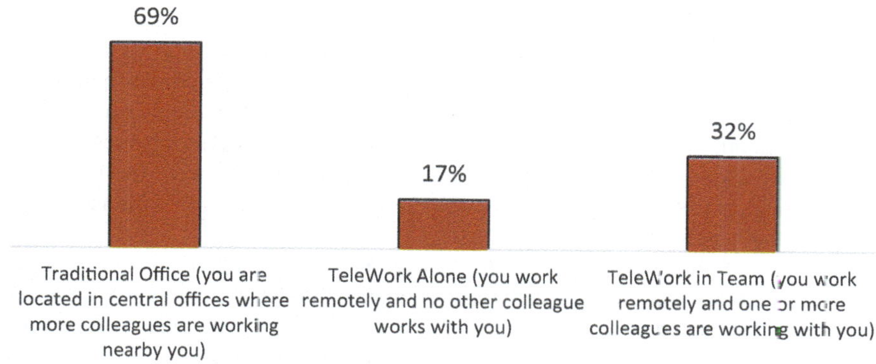

Fig. 2.10 Participants' office setting

the fact that most of them benefited from the materials and how they were organized to accommodate geographically decentralized remote teams. Another positive indicator was that over 70% rated the course understandability between 4 and 5. This reflects well on the format of course as a viable model to replicate among international "remote teams". Half of the participants rated the course difficulty at 3 out of 5. This suggests that the course was appropriately challenging and allowed participants to apply themselves.

With regards to rating the length of the course, opinions diverged as illustrated by Fig. 2.11. 36% of the participants deemed it at an appropriate length but at the same time an equally weighted group thought it was an inappropriate length. This finding suggests that for some people, the four-week period was sufficient to allow teams to form and to reach an optimal point to produce high-quality output. On the

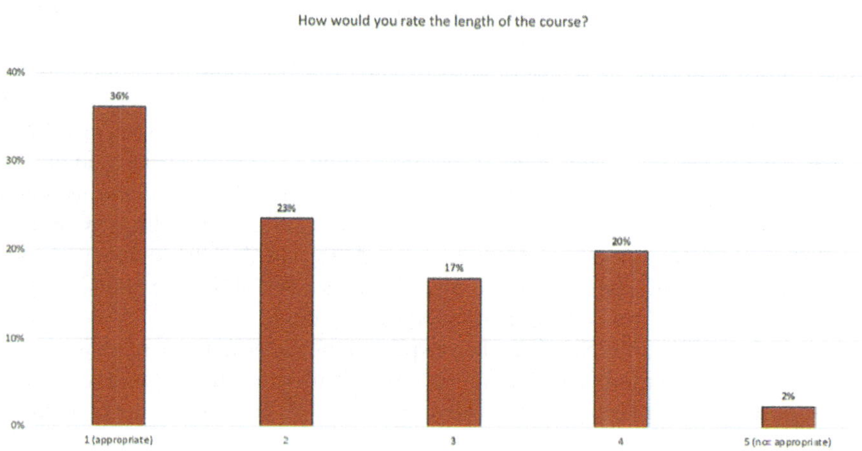

Fig. 2.11 Rating the length of the course

flip side, for others the time was insufficient. This shows that flexibility in time or allowing certain teams to work at their own pace can potentially improve the overall "remote teamwork" experience. However, regardless of having some difficulties in completing the course, almost all the participants said that they would recommend the course to others. Moreover, over 70% supported the idea of offering a second part of the course. This is a very good indication of the participants' perceived impact of the course, in addition to a promising finding for further spread of the skills associated with "remote teamwork".

When asked about the extent to which the course content will be usable after the completion of the module, almost all the participants rated it at more than 5 out of 10. This points to a high degree of transferability of the content to other fields outside the course. More specifically, the participants pointed to the reflection on poor teamwork experiences, as well as to the final reflection on the overall teamwork interactions during the course, as potentially guiding them on what they should/shouldn't do in teamwork in the future.

The HPI course instructors recorded 32 videos of about 5 min each, which presented the learning content during the first three weeks of the course. The videos were most of the time a dynamic discussion of two members of the team debating about a specific topic. The layout of the videos was designed to have the instructors in one half of the screen, and slides of a presentation outlining the keywords of the discussion in the other half of the screen. The slides were immediately available to the participant in a pdf format for easy download. In week 1, the different "Online Collaboration Tools" available at the moment of recording available were explained. In week 2, the discussion was centered on "The Human Side of Remote Work", and in week 3, the importance of "Intercultural Competences"—which has not been explored widely in the context of remote teamwork in the past- was introduced. After each video, the participants were asked to respond to some questions and they immediately had the chance to validate if their answers were correct or not, finding also the respective explanation of each answer. The HPI Team encouraged the participants at the end of each video to share on the collaboration space their experiences and opinions regarding the topics discussed. The team was moderating most of the topics suggested by the participants, and also provided additional information not covered in the videos such as published researches, ongoing studies, and also links to different sources that could support the discussions. The participants had all the time the availability to access all digital material as well collaborates via web and via App for iOS and Android. Tables 2.4 and 2.5 summarize the findings generated from the open reflection exercise of "I like…" and "I wish…". This section provides deeper insights on the elements that the participants liked about the course versus the elements that they wished were provided by the course.

Although several participants met their expectations and asked for a second edition of the course, some participants required more clear rules to manage discussions and how to perform team tasks more effectively, as well as to receive more practical studies as reference. A consolidated list of references to review the whole materials discussed would be also very useful to many participants. On the other hand, participants were satisfied with the structure, content, collaborative approach, learning

Table 2.4 Outlining the course elements that the participants liked

I like…
– Being part of a diverse group of participants
– The platform offered for this course
– The HPI Team
– The arrangement of the videos
– The content was interesting
– The teams' collaborative approach to solving technical issues
– The evaluation design of the practical tasks
– The Ease of teaming up with total strangers
– Tackling a complex assignment as a team
– Learning from the diverse modes of operation
– Learning from another teams' output
– The length of the videos was just right
– The questions after each video were very helpful
– The opportunity for self-improvement
– Availability and responsiveness of instructors

from other's experiences, self-improvement, and having the feeling of being part of a diverse group of participants with common interests. The expectations about the content were achieved, however, coordinating participation and making consolidated references available would have been a plus for those participants that could not take part continuously in the course.

Reflections on the Teamwork Experience

A deeper look at the team assignment shows that a little less than 50% of the participants said that the team experience was great and that they are looking forward to take part in further courses that include elements of "remote teams". It is also worth mentioning that 14% of the participants deemed the course as a good experience but nevertheless, they would not want to participate in similar courses. Whereas 21% did not have a good experience, which they attributed to having a bad team constellation. This highlights a clear variation in the team assignment experience, which is analyzed more closely in the following paragraphs.

Reflecting on the participants' decision to do the team assignment highlights that only 29% of the participants wanted to do it regardless of the nature of the specific challenge. This reflects a strong intrinsic commitment/desire to participating in teamwork. However, 44% of the participants were motivated to participate in the team assignment after reading about the challenge. The rest of participants said that they did not wish to participant in the team assignment. This finding suggests that some people are just team players, while others need to be persuaded by providing them with the appropriate challenge and conducive environment to apply themselves in a team-setting.

Table 2.5 Outlining the participants' wishes

I wish…
– Fewer videos
– More text-based documents
– Mobile phone access to the course content[a]
– Slides with focused and structured content
– A shorter course
– Better teamwork experience
– More support for the participants with limited technical knowledge
– A single file combining all the weekly slides in pdf format
– List of relevant literature
– Forming teams based on expectations with regards to what they want to do and their working style
– More time to work on the teamwork assignment
– Better explanation of the team tasks
– More focus on the personal and social aspects of remote teamwork and then decide which technological tools are appropriate
– Rules to govern the discussion forum
– Fair grading scheme that is less reliant on peer reviewers
– Better App compatibility with IOS
– More practical case studies

[a]Most of the course content is accessible on mobile phones either by following the course in a mobile browser (the platform has a responsive design) or by using the native mobile apps. Some features, e.g. the Collab Spaces are not available in the native apps

The participants' evaluation of the implementation phase of the teamwork was also reflective of an overall positive experience. Looking at one of the early tasks which is creating the team profile, 45% said that it facilitated their work on tackling the different components of the challenge. Another 33% associated it with learning about diversity, which they deemed as helpful. However, the rest (17%) evaluated this part of the team assignment as unhelpful.

Figure 2.12 shows that participants had varying experiences with self-organization and coordination within the teamwork. On the one hand, 29% of the participants had problems at the beginning but they were able to overcome these challenges at the end. This is an expected finding as the team members usually take time to get to know each other and reach a harmonious work relationship. On the other hand, around 24% of the participants experienced very good organization and coordination with their teams. This finding is likely representative of the participants who enjoy working in teams. 12% of the participants found it very difficult to get into groups. Though it is not a big percentage, it is still indicative of limitations in the problem-solving mechanisms in

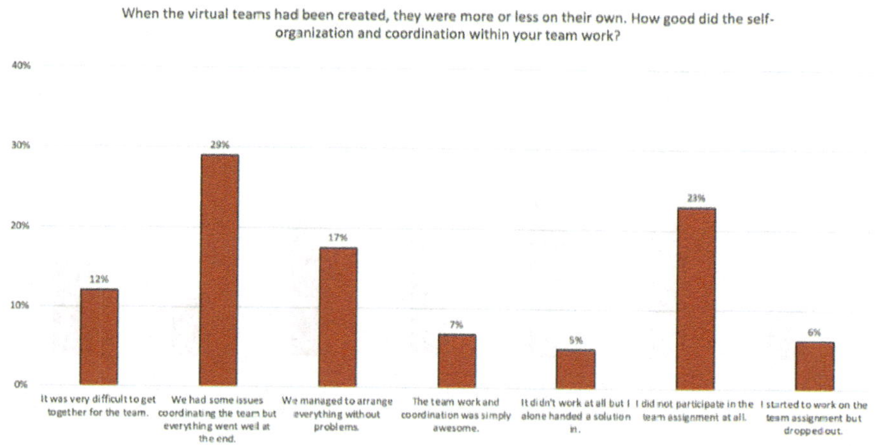

Fig. 2.12 Reflecting on the self-organization and coordination within the teamwork

the course, as those people should have been adequately supported. A small minority of 5% had a very bad experience with self-organization and coordination, but they still managed to produce a finished product.

Self-organization, and coordination within the teamwork, which Willcoxson's settings did not require, were strategically designed in the course to identify how teams were able to come along, and to find out if teammates were able to identify their leadership roles. The previous task, where teammates were requested to know each other by filling the team profile matrix, detailing the team members' pseudonyms, age, gender, expertise and supporting skills, meeting time preferences, leadership preferences, technological resources available and previous experience, was however, essential to make the blind-remote-interaction effective. The positive experiences reached 57% and no leadership conflicts were reported.

Reflections on the Communication Tools

The participants had a wide variety of communication tools at their disposal throughout the duration of the course. The teams decided by themselves which tools to use based on their expertise, time-zone locations or availability of access. Synchronous communication tools as well as asynchronous communication tools were proposed during the course. However, the participants chose to focus on some, more than others. The only tool that was required for the participants to use was the 'openHPI discussion forum' for dialogues where the course instructors were present answering questions of the participants and also providing supporting material. Figure 2.13 summarizes all communication tools presented and highlights the tools that were used the most by participants. It shows that almost half of the participants preferred to use the discussion forum in the Collab Space. The next most used tool was the Jitsi Meet which is also located in the Collab Space. To a lesser extent, the participants utilized tools like text chat, WhatsApp, and alternative video chat

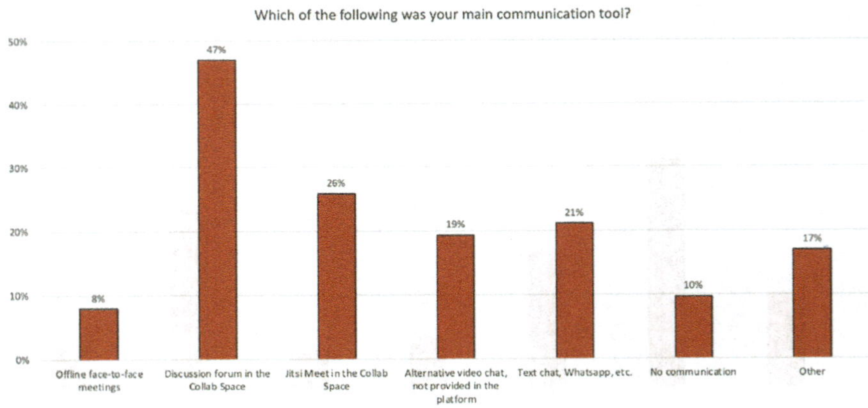

Fig. 2.13 Main communication tools used by the participants

applications not provided by the Collab Space. Assessing the functionality of the Collab Space showed that the file sharing feature is the one that is in need of urgent improvement. Other aspects were also deemed as in need of improvement: team forum (16%), video chat (14%), Etherpad (13%), and peer assessment tool (7%).

A closely related issue is the concern over the security of these online communication tools. The participants' thoughts on the issue show that more than half of them were not concerned about the security during their interaction with the team members. Around 21% expressed concern but they were still open to say whatever they wanted to say without self-censorship. This is a positive finding as it points to a trust in the tools, which are necessary to operate in "remote teams".

Participants were requested to use openHPI internal Collab Space for discussions and this can explain why the forum had a higher use and appears as one of the preferred tools. However, other asynchronous text-based tools such as social messaging and text chat (21%) did have also wide acceptance. Synchronous communications like face-to-face, Jitsi, and video chat for teams' discussions were not offered by the openHPI platform. However, many teams decided to use them with a considerable high acceptance (53%).

Reasons for Dropping Out of the Team Assignment

When asking those who partially took part in the team assignment about the reason why they dropped out, the most common reason was the lack of time due to unforeseeable personal developments, and as a result, not being able to find an appropriate timeslot to meet with the other teammates. A very good indicator of the functionality of the team-building criteria/process is that only a negligible minority dropped out of the team assignment as a result of not getting along with the team members. Previous studies also showed that lower performing participants are very likely to drop out from the teams [31].

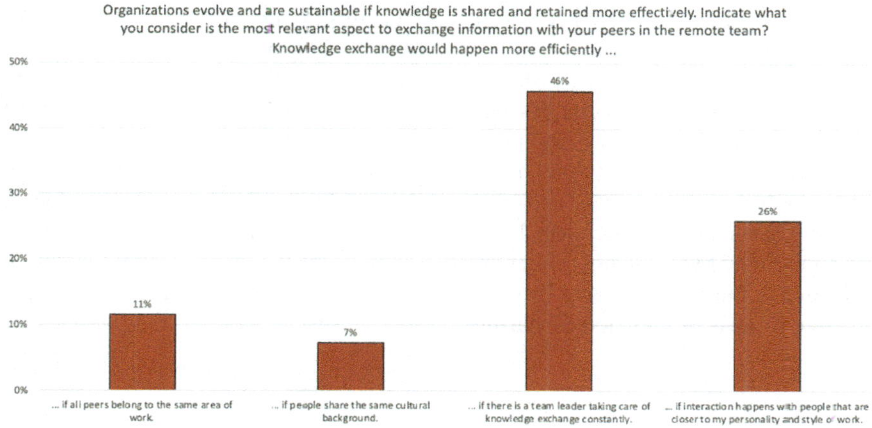

Fig. 2.14 Ways to improve knowledge exchange among remote teams

Ways to Improve Teamwork in "Remote Teams"
Figure 2.14 presents the participants' thoughts on the ways to improve knowledge exchange among "remote teams". As illustrated below, almost half of the participants attributed the effectiveness of knowledge exchange to having a good team leader. A less popular thought is the association of effective knowledge exchange with the team members sharing similar personalities and styles of work. This is an interesting finding, which suggests that external factors like time zones are more influential on team dynamics than internal/intrinsic factors like personalities and work style. This is a positive finding as it implies that external factors can be used to optimize teamwork.

2.4 Conclusion and Future Work

In this study, we showed that assessment of digital remote teams in online training environments can be successfully implemented by means of assisting team building and encouraging virtual participation.

The concept suggested by Willcoxson to develop and assess project management and resourcing strategies, concurrently with team dynamics instead of sequentially, requires in the remote environments the definition of team-member profiles and their roles within the remote teams. It does clearly encourage high responsiveness of all members and also allows to accurately assess individuals' contributions to teams, and also project planning and resourcing.

The definition of a macro project plan from the very beginning of the course with clear tasks, goals, and dates helped the teams to achieve the targets of the course on time. It also allowed remote peers to plan in advance, not only their own work, but also to coordinate the work that was expected to be done as a team.

We also identified how the freedom of the team to select their own synchronous and asynchronous online tools enables the necessary transparency during interpersonal interactions, independently of the geographic location or time zone.

Future work should explore how to combine local and remote teams using online tools to communicate and compare how collaboration may change due to the fact that team-members and leadership roles have different personal interactions within the hybrid teams. Other areas to explore are the internal communication dynamics of culturally diverse "remote teams" and how it hinders and/or enhances the effectiveness of knowledge sharing.

References

1. Riebe, L., Girardi, A., & Whitsed, C. (2016). A systematic literature review of teamwork pedagogy in higher education. *Small Group Research, 47*(6), 619–664.
2. Staubitz, T., & Meinel, C. (2018). Collaborative learning in MOOCs—Approaches and experiments. In *Proceedings of the 48th IEEE Frontiers in Education Conference (FIE)* (pp. 1–9). San Jose, CA, USA: IEEE. https://doi.org/10.1109/FIE.2018.8659340.
3. Mark, G. (2001). Meeting current challenges for virtually collocated teams: Participation, culture, and integration. In L. Chidambaram & I. Zigurs (Eds.), *Our virtual world: The transformation of work, play and life via technology* (pp. 74–93). Hershey, USA: Idea Group Publishing.
4. Jarvenpaa, S. L., & Leidner, D. E. (1999). Communication and trust in global virtual teams. *Organization Science, 10*(6), 791–815.
5. Ahuja, M. K., Galletta, D. F., & Carley, K. M. (2003). Individual centrality and performance in virtual R&D groups: An empirical study. *Management Science, 49*(1), 21–38.
6. Paul, D. L., & McDaniel, R. R., Jr. (2004). A field study of the effect of interpersonal trust on virtual collaborative relationship performance. *MIS Quarterly, 28*(2), 183.
7. Gits, T. R. M. (1992). Self-efficacy: A theoretical analysis of its determinants and malleability. *Academy of Management Review, 17*(2), 183–211.
8. Malhotra, A., & Majchrzak, A. (2004). Enabling knowledge creation in far-flung teams: Best practices for IT support and knowledge sharing. *Journal of Knowledge Management, 8,* 75–88.
9. Townsend, A. M., DeMarie, S. M., & Hendrickson, A. R. (1998). Virtual teams: Technology and the workplace of the future. *Academy of Management Executive, 12,* 17–29.
10. Fischer, J., Gündling, N., Harcks, A., & Schnöger, C. (2012) *Work@Home, Ein Kommunikationsprojekt, Gesellschafts- und Wirschaftskommunikation.* Universität der Künste, Berlin.
11. Jehng, J. J. (1997). The psycho-social processes and cognitive effects of peer-based collaborative interactions with computers. *Journal of Educational Computing Research, 17,* 19–46.
12. Civin, M. A. (1999). On the vicissitudes of cyberspace as potential space. *Human Relations, 52,* 485–506.
13. Fusaro, B. (1997). How do we set up a telecommuting program that really works? *PC World, 15,* 238–247.
14. Wellman, B., Salaff, J., Dimitrova, D., Garton, L., Gulia, M., & Haythornthwaite, C. (1996). Computer networks as social networks: Collaborative work, telework, and virtual community. *Annual Review of Sociology, 22,* 213–238.
15. Duxbury, L. (1999). An empirical evaluation of the impact of telecommuting on intra-organizational communication. *Journal of Engineering and Technology Management, 16,* 1–28.

16. Cerulo, K. A. (1997). Reframing sociological concepts for a brave new (virtual?) world. *Sociological Inquiry, 67,* 48–58.
17. Ngwenyama, O. K., & Lee, A. S. (1997). Communication richness in electronic mail: Critical social theory and the contextuality of meaning. *MIS Quarterly*, 145–166.
18. Workman, M. (2001). Collectivism, individualism, and cohesion in a team-based occupation. *Journal of Vocational Behavior, 58,* 82–97.
19. Atkinson, S. (1998). Cognitive style in the context of design and technology project work. *Educational Psychology, 18,* 183–194.
20. Lim, K. H., & Benbasat, I. (2000). The effect of multimedia on perceived equivocality and perceived usefulness of information systems. *MIS Quarterly, 24,* 449–471.
21. Fussell, S. R., & Benimoff, I. (1995). Social and cognitive processes in interpersonal communication: Implications for advanced telecommunications technologies. *Human Factors, 37,* 228–250.
22. Fritz, M. B. W., Narasimhan, S., & Rhee, H. S. (1996). The impact of remote work on informal organizational communication. In *Telecommuting? 1996 Proceedings* (pp. 1–20). Jacksonville, FL, April 1996.
23. Heald, M. R., Contractor, N. S., Koehly, L. M., & Wasserman, S. (1998). Formal and emergent predictors of coworkers? Perceptual congruence on an organization's social structure. *Human Communication Research, 24,* 536–563.
24. Chmiel, A., Sienkiewicz, J., Thelwall, M., Paltoglou, G., Buckley, K., Kappas, A., et al. (2011). Collective emotions online and their influence on community life. *PLoS ONE, 6*(7), e22207.
25. McGrath, J. E.. (1984). *Groups: Interaction and performance.* Prentice-Hall.
26. Chujfi, S., & Meinel, C. (2015). Patterns to explore cognitive preferences and potential collective intelligence empathy for processing knowledge in virtual settings. *Springer Open. Journal of Interaction Science, 3,* 5.
27. Willcoxson, L. E. (2005). Its not fair: Assessing the dynamics and resourcing of teamwork. *Journal of Management Education, 30,* 798–808.
28. Alario-Hoyos, C., Pérez-Sanagustin, M., Delgado-Kloos, C., Parada, A. G. H., Muñoz-Organero, M., Rodriguez-de-las Heras, A. (2013). Analysing the impact of built-in and external social tools in a MOOC on educational technologies. In D. Hernandez-Leo, T. Ley, R. Klamma, & A. Harrer (Eds.), *Scaling up learning for sustained impact* (pp. 5–18). Berlin, Heidelberg, Germany: Springer.
29. Grella, C. T., Thomas, S., & Christoph, M. (2019). Performance of men and women in graded team assignments in MOOCs. In *IEEE Learning with MOOCs Conference (LWMOOCs)*. IEEE.
30. Staubitz, T., Hanadi, T., & Christoph, M. (2018). Team-based assignments in MOOCs—user feedback. In *Proceedings of the 2018 IEEE Learning with MOOCs Conference (LWMOOCS)* (pp. 39–42). IEEE. https://doi.org/10.1109/LWMOOCS.2018.8534607.
31. Staubitz, T., & Christoph, M. (2019). Graded team assignments in MOOCs—Effects of team composition and further factors on team dropout rates and performance, In *Proceedings of the Sixth Annual ACM Conference on Learning at Scale (L@S)*. Chicago, IL, USA: ACM. https://doi.org/10.1145/3330430.3333619.
32. Payne, H. J. (2005). Reconceptualizing social skills in organizations: Exploring the relationship between communication competence, job performance, and supervisory roles. *Journal of Leadership and Organizational Studies, 11*(2), 63.
33. Hughes, R. L., & Jones, S. K. (2011). Developing and assessing college student teamwork skills. *New Directions for Institutional Research, 149,* 53–64.

Chapter 3
Collaborative Work in Higher Education: Tools and Strategies to Implement the E-Assessment

M. P. Prendes-Espinosa, I. Gutiérrez-Porlán, and P. A. García-Tudela

Abstract This chapter is an educational approach to online collaborative work in higher education, considering both the digital tools and the strategies for the e-assessment. We can observe the evolution of digital learning: from e-learning to m-learning and u-learning, through blended-learning and new other models like adaptive learning. In every model, we can try to promote collaborative work and we need to assess these group strategies. Thus, our main objectives are answering what type of digital tools we can use to promote virtual collaboration, which are the main strategies to implement, and finally considering different models of e-assessment in relation to the previous issues. Educational research shows the possibilities of digital tools like social networks, e-portfolios, collaborative environments, or traditional tools inside LMS from an assessment approach that considers the formative processes as the key to online collaboration.

Keywords Online collaboration · E-assessment · Web 2.0 · Teaching and learning strategies · University

3.1 Educational Approach to Collaborative Work

Collaborative methods are really relevant in e-learning, because these methods have meant the evolution from one-to-one relations to group relations as the base of learning in distance education. Collaboration is a way to learn and it is also a methodology to teach, both in face-to-face and virtual educational situations. As

---

M. P. Prendes-Espinosa (✉) · I. Gutiérrez-Porlán · P. A. García-Tudela
Department of Didactic and School Organization, University of Murcia, C/Campus Universitario, 12, 30100 Murcia, Spain
e-mail: pazprend@um.es

I. Gutiérrez-Porlán
e-mail: isabelgp@um.es

P. A. García-Tudela
e-mail: pedroantonio.garcia4@um.es

R. Babo et al. (eds.), *Workgroups eAssessment: Planning, Implementing and Analysing Frameworks*, Intelligent Systems Reference Library 199,
https://doi.org/10.1007/978-981-15-9908-8_3

[34] acknowledge "advanced understanding of suitable pedagogical design principles and knowledge from empirical cases is needed" (p. 2). We agree with them that collaborative skills are crucial for the professional future of our university students. So, these are our main objectives: first, to analyze some principles for the instructional design of virtual collaboration in higher education; second, to analyze the possibilities that some digital tools offer in the development of the methodology and the e-assessment; third, to explain our approach to e-assessment based on this pedagogical and technological knowledge; finally, to present practical recommendations for instructional designers and teachers. This knowledge is the result of our experience of more than twenty years teaching online with collaborative methodologies and also our collected data from research.

To understand our current situation, it could be useful to look back in history. When the Internet was born, the beginning of teaching and learning processes relying on new technologies was based on the access to digital contents and also in the action of teachers and tutors. Thus, the cognitive interaction was always between one student with his/her teacher and the instrumental interaction, supported by non-interactive contents.

At the same time, the models of collaborative work in classrooms were designed for traditional learning (face-to-face) in the 90s. This innovative proposal was conceived to address students' isolation in the class and to challenge the teachers' usual strategy of promoting competition between those students. In the field of compulsory education, Johnson & Johnson developed a model of learning based on interaction among students, self-responsibility (both individual and group) in the achievement of the task, and positive interdependence. This model was designed for face-to-face learning, but its theoretical principles are the basis of virtual collaboration as well as e-learning methodology in the tradition of CSCL (Computer-Supported Collaborative Learning).

Thus, in the following years and towards the end of the last century, the first phases of e-learning were being replaced by other advanced methodologies based on group and collaborative interaction, since we understood the relevance of interaction using telematic tools to promote significant learning in distance-learning processes [25, 46]. So, in those years we adapted face-to-face collaborative techniques to virtual environments. This is the same decade where we have seen the big growth of the Internet with the development of World Wide Web and applications like Hotmail or Mosaic.

In the current century and in these changing times, we have seen the development of Web 2.0, social networks, Facebook, YouTube, Google, etc. The evolution of technologies gives us the opportunity to develop new models in digital education like blended-learning, mobile-learning, or ubiquitous-learning. These models entail the revolution of new online education, the access to education from anywhere and by anyone. So, we developed the learning communities, open education, massive courses, online collaboration, etc. a complete catalogue of new proposals to innovate in education, above all when we are talking about adolescents and adults. Different approaches must be implemented when dealing with children, but this is not our current objective.

As we have seen, this historical approach gives us the idea of the parallel evolution of internet and e-learning models. In this way, we can talk about the evolution of education based on collaboration, because we began by adapting the models created for face-to-face education, but at the present time, we have our own collaborative models creating virtual communities and developing the PLN of students

Nowadays, we cannot talk about e-learning models if we only consider the isolated student in their room. Today's student is always connected to a world of contents and of people interconnected like him, too. For that reason, we use the concepts of PLE (Personal Learning Environment) and PLN (Personal Learning Network) to describe the new technological learning environments where the student has plenty of opportunities to learn, thanks to the connections between people and to open digital resources.

With regard to these topics, our research group finished a quantitative study in 2018 about the PLE and PLN of university students with 2054 answers from all the Spanish universities [69]. Our main research problem was the next: the type of digital tools and online strategies that our students use to communicate and collaborate with their peers in the academic work. The results show that in general, students prefer messaging apps (41%), followed by email (28%) and then social networking services (26%) for the academic tasks when they need to communicate with others. Less than 6% opted for video-conference tools. When students were asked about their preferences regarding the tools used to carry out group work or collaborative work, the majority of them use Google Drive (64.5%) followed by social networks (22.35%), virtual environments of their universities (7.98%); however, they did not even consider other tools like wikis or blogs. Moreover, students were asked about aspects they prioritize when working in teams. The majority of the students considered as priorities the fact of "building knowledge together" (58%), "interacting with others" (53%), and "sharing resources" (49%).

To understand the present time, it is also relevant to mention the "open" movement: open science, open education, open courses, open access…. In education, this openness is linked to collaboration, because the professional communities of teachers are responsible for the huge number of open webs and open educational resources. These teachers and education professionals have understood that collaboration can be the best way to guarantee the future and the social impact of our work.

In summary, we are going to analyze in this chapter the possibilities of collaboration in higher education and we will use some real examples. This analysis will be useful to understand the models of e-assessment when we have to work in these virtual environments and with collaborative strategies to teach and learn. And we will suggest some current telematic tools that can be useful to develop a virtual collaboration and to assess these processes, too. To help you read and understand this work, here are some of the main concepts used throughout the chapter:

- *Virtual collaboration* is a specific model of sharing and cooperating, developed on the net, and supported on both synchronous and asynchronous telematic tools. From an educational point of view, it is a methodology to teach and a strategy to learn. It is very usual in e-learning models to promote more interactive processes.

- We understand *social networking tools* as those services or applications characterized by allowing the user to create a data profile about himself/herself online and share it with other users.
- *Collaborative Workspace* tools are a group of tools that have been specifically created for designing and implementing collaborative processes.
- *LMS (Learning Management System)* is software installed on a web server that is used to manage, distribute, and control the non-face-to-face training (or online education) activities of an institution or organization. They are known as Virtual Campus platforms at university level.
- *Web 2.0 tools* are a group of digital and online tools in which users play the principal role in the communicative process, thanks to a technological simplification which allows a greater number of users to carry out actions which only a few could perform before.
- With *social media* sites we refer to online environments where users publish and share digital objects around which a network of people is generated. The base of the collaboration lies on the shared object (videos, presentations, bookmarks, files, etc.).
- In relation to the *4 types of e-assessment* discussed in this work, it should be noted that two of them are usually led by the teacher, although their development is very different: *e-formative* assessment consists of constant and immediate feedback through social networks, blogs, etc., in order to optimize the teaching and learning processes. On the other hand, *e-summative* assessment consists of the assessment applied at the end of a task, lesson, course, etc. Tests or interviews through videoconferences are usually used.

Students are the main protagonist of the two remaining types of e-assessment. One of them is *e-self assessment*, from which each student must develop a reflective process to identify skills and strengths in relation to the objectives being evaluated. Generally, digital tools based on collaborative e-portfolio are usually used. Finally, *e-peer assessment* consists of evaluating the results and/or processes of a couple or a group. In addition, this type of assessment is part of the paradigm of learning by teaching and various resources are often used (cloud office tools, social media, social networks, etc.).

3.2 Collaborative Techniques in E-Learning

Telematic networks, for their own nature, make possible the collaboration between users. Online collaboration in education is an interesting educational strategy which allows us to harness the benefits of collaboration combined with the benefits of digital technologies. As we have explained, the origin of these virtual strategies is in face-to-face collaborative models, which we have adapted to the online process of teaching and learning [46]. We consider three phases: instructional design, implementation, and assessment; the latter does not only take place in the end, it is carried out from the beginning (the decisions about the procedures of assessment), during the implementation (collect data, observe process of collaboration) and it is also the last task for teachers. We have been working in the designing of this model along more than twenty years teaching with collaborative methods to postgraduate university students. The first version of the model was described in [67], but we made changes and we improved the proposal during these years. The last sequencing of our model is represented in Fig. 3.1.

In this figure, we can observe that we have teachers who have a very relevant role as designers, but they also have to be present during the implementation. And the assessment is in relation to both processes (collaboration itself) and final products (results of tasks).

We can find a lot of manuals and articles about collaborative techniques in relation to the age of students, to the objectives or other relevant dimensions. As a starting point, we use the result of the European project CARMA [70] and we have selected the most interesting collaborative techniques used in this project. In this research, we demonstrated that not only are these techniques useful to learn but also motivating students when learning [34] and therefore contribute to prevent academic failure and early school leaving [70] We have used these techniques of collaborative learning both in face-to-face and virtual university teaching and we have chosen two of them

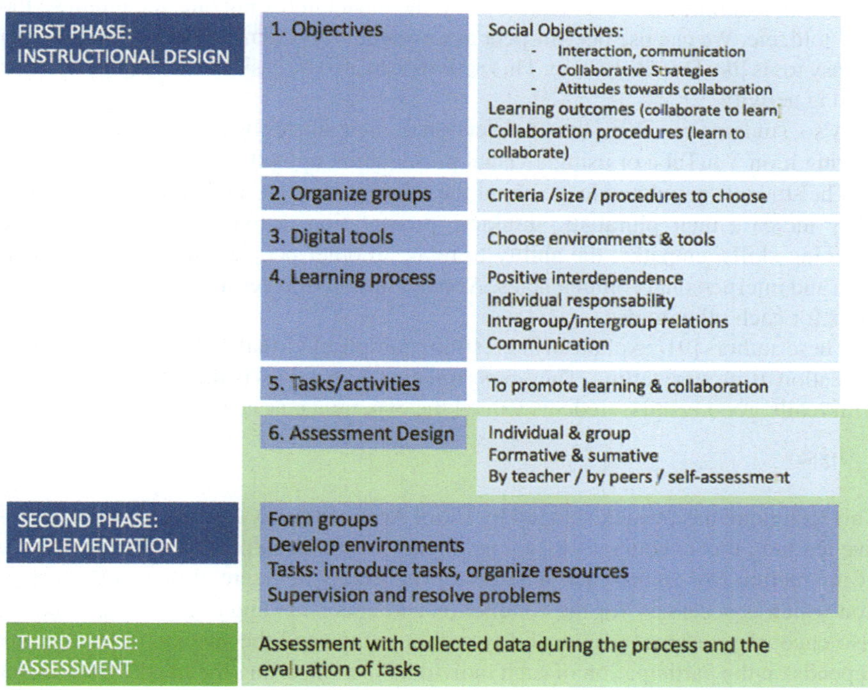

Fig. 3.1 Model of online collaboration by Martínez and Prendes [46]

with proven optimal results with university students working online, according to our research [70] and other similar works [91, 93].

Both techniques enhance this chapter with concrete examples of how to bring collaborative work and e-assessment into the university classroom.

- Learning Through Storytelling

Learning through storytelling is a process in which learning is structured around a narrative or story as a means of "sense making". It involves the use of personal stories and anecdotes to engage students and share knowledge. The aim of the activity is to create a story together on a topic of interest, it can be done in small or big groups.

One application of this technique is to start sharing a story grid with the students with different words or concepts relevant for the activity. The story grid can be of any size, but usually, the bigger the grid, the more complicated the activity will become. Students can use any vocabulary or grammar they want to, but they have to include all the words in the story grid. There are different telematic tools to develop this phase: for example, the story grid can be published on a wiki where students can create the story together easily, the majority of LMS have wiki tools, for example, Moodle or Sakai. They can, alternatively, use a digital tool to build conceptual maps like Cmaptools, Bubbl or Mindmeister.

At the end of the activity, students could vote for the best stories in different categories, for example the most creative story, the most interesting one, the funniest, the best-told, etc. We can use peer-to-peer assessment tools (some tools on TeachAssist) or easy tools like Doodle to vote. This activity can also be easily turned into a creative writing activity.

Also, students can share the final version of their stories by recording a video and sharing it on YouTube or using a visual presentation uploaded on Slideshare.

The students gain knowledge about the relevant topic through a new perspective. They increase their pluralistic thinking, presentation, active listening and public speaking skills, empathy, the ability to relate to other people, and strengthen their intra and interpersonal competencies. Another option is to get students to create story grids for each other to use.

These authors [91] explain an interesting example of virtual collaboration in higher education using storytelling. They experimented with videos in the form of a narrative with really good results, students valued the experience as new and inspiring.

- Jigsaw

This technique has been developed by Elliot Aronson and it is a method of cooperative learning that encourages listening, engagement and emphasizes the importance of interaction (by giving each member of the group an essential piece of information which is necessary for the completion and understanding of all the material). It also encourages shared responsibility within groups and the success of each group depends on the participation of each individual in completing their task. The teacher has to select the topic, to introduce it to the students, and assign students into heterogeneous "home base groups" (4–5 students per group). After it is necessary to divide the material needed to cover the topic (articles, reports, problems, etc.) into segments (as many pieces as number of group members). Assign each learner to learn only one of these segments. Each member must learn the material pertaining to their section and be prepared to discuss it with their classmates. Teachers must give students enough time to read and learn their segment and become familiar with it. We can use virtual folders on the cloud or use email to organize this phase.

After, the teacher has to form "experts' groups": once students have learned their part, they move into expert groups by having one learner from each home base group join other students assigned to the same segment. Explain them that they have to share ideas and discuss the main points of their segment and plan how to present the information to their home base groups. The teacher will give students in these expert groups time to discuss the main points of their part, and to prepare and rehearse the presentations they will make to their home base group.

Next, students have to return to their home base groups and take turns teaching their area of expertise to the other group member, so each home group will have information about all topics. Ask each learner to present her or his segment to the group. Encourage others in the group to ask questions for clarification. At the end of the session, give a quiz on the material. At this time, team members should not help each other.

In these phases the communication is the most important strategy, so we can use forums, chats, videoconference, etc., any communication tool; thus we can combine asynchronous and synchronous strategies. From our experience, the most accurate telematic tool to develop this technique is a collaborative platform in which teachers can develop different spaces for the discussion and folders with resources for every group, as for instance BSCW (we have tools to share resources and to promote the discussion between students). We can also use virtual folders as Dropbox or any other tool on the cloud to organize resources and groups, for instance, Google Drive or Edmodo. Another possibility is to search a LMS with group tools, like Sakai.

Vargas-Vargas et al. [93] developed a course with mathematical content in which they introduced Jigsaw within a virtual learning environment. Using this technique in a virtual context allowed students to benefit from collaboration and e-learning. According to the authors, students who have participated in the experience have a greater identification with the study of statistics, show improvement in their attitude towards statistics, and an overall better learning process, reporting improvements in the evaluation process.

3.3 Possibilities of Telematics Tools

In the previous sections, we have pointed at the importance of the collaboration in education and its advantages for the enhancement of the students' learning process. When we collaborate with others, we force ourselves to interact; being the responsibility of everyone for their own and others' learning a key element in this process [88].

Table 3.1 Recommendations to choose the digital tools to collaborate

Technical	Infrastructure and communication facilities at the workplace
Ubiquitous	Possibility of working with the tool anywhere and anytime and from any device
Audience	The difficulties of using the tool and the level of skill that users have are a key aspect of the fine development of the collaborative process. The use of tools should ease our lives and not make them more complicated. The focus must be on the learning and teaching, not on the tool
Purpose	We have to consider the tools' ultimate purpose. There are many instruments with which we can perform a great variety of tasks, but it is always more convenient to use each tool for its own purpose; that is, if we need to communicate synchronously, a videoconference or a chatroom are going to be the finest options

While it is a fact that collaborating with others is a great opportunity, in numerous occasions, the time–space limitations render the collaborative processes impossible. In this sense, technologies represent a necessary response for surmounting said limitations and as an efficient solution for the evaluation process. We have more and more chances of not only having access to others' ideas or work, but also of getting other involved in our own knowledge-building processes, by means of virtual environments, sites, and tools which facilitate our collaboration with them [47]. These tools provide common spaces to carry out tasks collaboratively, which allow us—apart from performing the task itself,—to learn together, even when we cannot share the same physical space (for time or distance reasons).

The possibility of long-distance communication and work with them has been and still is one of the technological advances which have had a greater impact along history of humankind [89].

The range of tools that are part of Web 2.0 is really wide and many of them can be employed for online collaboration and evaluation.

When choosing the most appropriate tool and using it in the most optimal manner, it is important to bear in mind a number of recommendations (see Table 3.1).

Next, we feature different telematic tools which enable online collaboration and the evaluation processes linked to this collaboration. We are going to present them into three groups: Learning Management Systems (LMS), Collaborative Work-space (specific tools for collaboration), and web 2.0 tools.

3.3.1 Learning Management Systems

Just like there are telematic tools whose specific use is that of collaboration, as we have seen in the previous section, the incorporation of technologies in teaching–learning processes has brought about the creation of tools specifically aimed at the implementation of these. They are known as LMS (Learning Management Systems) or Virtual Campus platforms, a name given by universities in an attempt of these to create a university campus in the scope of virtuality which enables the students to

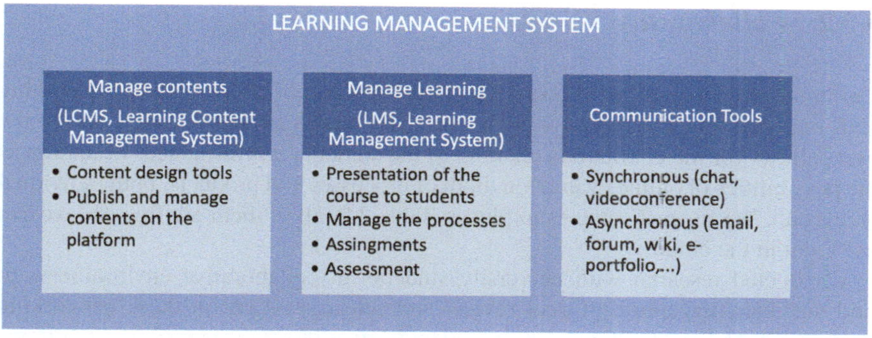

Fig. 3.2 Elements of learning management systems

access content teaching, its organization (classrooms, registration, etc.), and to the rest of complementary spaces such as the library, university services, etc.

These Learning Management Systems require the installation of a given software in a server and integrate into one same space the necessary tools for the management of the whole educational process. We can, specifically, highlight three functional elements [54], see Fig. 3.2.

The majority of the platforms currently in use, being free software or not, include all the described applications: LMS, LCMS, and some communication tools (many of them are known as web 2.0 tools aimed at improving communication and collaboration, which we will look at in more detail in Sect. 3.3). There is a multitude of virtual campus platforms, amongst which we can highlight some free software ones, such as Sakai, Moodle, Dokeos, Claroline, or LRN, for instance [34, 54] recognize that LMS are not always thought to promote collaboration, but it is possible to collaborate on them.

The possibilities of assessment within the framework of these platforms are quite numerous. On the one hand, we find tools specifically used for evaluation: an exam-designing tool, questionnaires, facial recognition tools for examinations by videoconference, among others. On the other hand, these tools allow the submission of assignments, both individual and group, monitoring the student's activity online and implementing peer-evaluation and self-evaluation in the assignments submitted.

In a study carried out with first-year psychology students at Strathclyde University by [35], it was found that the main advantage of evaluating network collaboration through LMS (Moodle) is the increase in motivation on the part of the students. Peer assessment throughout the course using Moodle led to a significant improvement in the final grade of students, especially those who were less motivated by the subject at the beginning. The authors highlight the advantages of this type of closed networked space for the control and monitoring of evaluation and peer-evaluation among students.

3.3.2 Collaborative Workspaces

Currently, there is a group of tools which have been specifically created for designing and implementing collaborative processes. The record of tasks carried out by the users allows for the continuous tracking of the students' performance, which opens the possibilities of online evaluation, both of processes and products, both individual and group. We suggest some examples in Table 3.2, all of them are free and we can use them in the cloud.

Conducted research with university students in collaborative environments to find out their perception of peer review and self-assessment in these networking spaces [38]. The results showed improved understanding of the perspective of others, improved self-awareness, and the creation of an online learning community that enhanced participation, evaluation processes generating a strong sense of community.

3.3.3 Tools from Web 2.0 for Collaboration

The term Web 2.0, used for the first time by Dinucci [16], gives us the notion that we are dealing with an evolution of the world wide web (WWW), a further step forward on the original web. The author of the term foresaw in 1999 that the web which allowed us to access information from our browsers would be the embryo of what was to come, defining it as a medium for interactivity. And this is precisely where we are today, forming part of a web where we users, are the actual protagonists.

We find in [59] one of the most accurate approaches in this regard. The author, along with his colleagues at O'Really Media referred to the term Web 2.0 during a lecture in which what has been known as the "*dotcom bubble burst*" was being debated. Analyzing the reasons why some of these companies (Google, Amazon, eBay…) had survived said burst [60] and in trying to understand the business model upon which those companies are grounded, the seeds of what we know today as Web 2.0 originated.

In the work of these authors, the Web 2.0 is understood as an evolution of the traditional Web, where users play the principal role in the communicative process, thanks to a technological simplification which allows a greater number of users to carry out actions which only a few could perform before. Users thus become both readers and writers and creators of the information existing on the Internet.

The Web 2.0 offers a wide range of possibilities regarding online collaboration and evaluation. Being part of this web means being part of a network of people who build, collaborate, and share by means of different tools and spaces in which one "is" [68].

We feature next some of the most remarkable tools of the Web 2.0, highlighting the possibilities of these for online collaboration and evaluation. It is relevant to consider that the majority of these tools are on the cloud and according to [54], peer-to-peer communication and collaboration using cloud-based-tools (CBTs) are valuable

Table 3.2 Tools for implementing collaborative processes

Origin	Access	Utilities
BSCW ("Basic support for cooperative work")		
Designed and manufactured by the FIT (Institute for the Application of Information Technology), a research unit of the GMD (German National Center for Information Technology Research)	From any devices, with version for Smartphone (iOS and Android) and the installation of desktop applications, which makes easier its access and use	– Store, share and manage files – Grant password-protected access – Manage appointments, contacts, tasks, and notes – Use versioning and change reports – Be updated on your teammates' activities – Blogs – Polls – Send automatic reminders
Trello		
Designed in 2011 by Fog Creek Sofware and it was sold to Atlassian in 2017	Access and synchronization of information from any device, since a mobile application for both iOS and Android are available	– To manage from one same work environment the communication and the creation and edition of any kind of document – Allows the integration on services and tools like Google Drive, Evernote, and Dropbox among others
Nuclino		
Created in Munich (Germany) by Björn Michelsen and Jonathan Kienzle	Mobile and desktop application for Windows, iOS, Android, and Linux	– Collaborative document editor – Agenda and meeting management – File browser – Synchronous and asynchronous communication tools – It allows a rather simple integration of other applications like Google Drive, Google Suite, Slack, and Google Drawings among many other
Slack		
Developed by Slack Technologies	Version for computer and usable on mobile devices. It has a version for business (not free)	– Easy integration with other services like Drive, Trello, Zoom, Calendar, and more – Share folders and files – Communication tools and organization of conversations – History of group work

(continued)

Table 3.2 (continued)

Origin	Access	Utilities
Microsoft teams		
By Microsoft	It has a free version and other versions for business (not free). You can download the application for desktop or for mobile devices	– Chat, videoconference – Share documents and folders – Collaboration on real time – Integration of other services on the cloud and with Microsoft Office

resources for promoting a strong motivation in university students. In general, they perceive these CBTs as a positive factor when they are part of the instructional design and part of the learning activities. The next classification of tools is made by [88].

- Wikis

The word wiki has its origin in a Hawaiian term which means "quick". This tool enables collaborative edition through the only use of a browser, can bear hyperlinks and its editor is usually a rather simple one. The most evident result is the well-known Wikipedia, an online encyclopedia created collaboratively between the people that have been selflessly introducing information.

Wikis can be implemented in higher education as a scope for the collaborative creation and publishing of a mutual product, for the creation of collaborative notes, for the publishing of activities in small and large groups, in a nutshell, they enable the creation of a common online environment where all individuals in a classroom have the same opportunities for information creation and edition.

All these possibilities pave the way for continuous assessment processes, since it is possible to track and monitor the tasks carried out by each student, the final assessment of the finished product and peer-to-peer evaluation, by being able to revise and modify what has been suggested by other classmates. This way the responsibility of everyone in the accomplishment of the task is fostered.

Some simple and free tools for the creation of wikis are PBworks or MediaWiki. Perhaps the most well-known example of wiki used as evaluation tool in collaboration is from [92]. In his experience, he uses wiki as a tool to promote co-writing in a group work, but also it is relevant that he proposes methods to evaluate each student's contribution to the collaborative process based on the automatic registered information by wiki plus survey grids.

- Blogs

A blog is a rather simple publishing tool which enables its author or authors to add content of interest in all types of format (text, video, etc.). The information published on a blog is presented in reverse chronological order and readers can make comments easily on each of the entries published. These comments can be subject to moderation or not, even in some cases comments can be disabled, being the blog used as a one-way communication channel.

The possibilities of this tool from an educational perspective are immense since its versatility makes it possible to introduce it in the teaching–learning processes in different ways [89]. In the context of collaborative work, blogs are an excellent medium for publishing content online within groups: progress of the work realized, information search for the completion of the activity, history of group activity along the work process, accounting for the distribution of tasks and the own collaborative process, task documentation or individual and group reflections on the progress of the task and learning processes.

The possibility for the teacher to access the information published and to comment on it permits a quick feedback and the opportunity for the student to make improvements on the contents posted. On the other hand, being able to access the content published by other classmates allows making comments on the posts and review and evaluate among peers the learnings accomplished. One of the most widely used and intuitive blogging tools is Blogger.

Among the main potentials for online evaluation, we can stress that it has become a useful tool of formative or summative evaluation, being one of the most used telematic tools for the university students' formative assessment [32]. Research in this field has shown that the use of blogs for formative assessment in higher education allows for the introduction of reflective processes of technology-enhanced peer learning. It also enables the improvement of the connection between new and previous learning without the direct involvement of a teacher through peer assessment and learning [62].

- Video-conferencing tools

Another example of widely spread collaborative applications is video-conferencing tools which enable synchronous communication with audio and video, such as the popular Skype, Google Hangouts, or Zoom. These tools facilitate and make possible the collaboration since they bring the users closer, by overcoming spatial barriers, allowing in some cases the use of shared work environments while communication is taking place.

In online evaluation, these tools become an indispensable virtual space for the presentation of tasks on e-learning models and also for oral exam testing. The study developed by [4] explored qualitatively the relationship between problem-based learning, authentic assessment and the role of collaboration in digital contexts based on video conferencing systems. They an increase in students' autonomy, the improvement of the participation and motivation, and the greater use of meaningful self and peer assessment. Moreover, authors found the improvement of collective knowledge through the synchrony of communication. The main conclusion was that this improvement lays the foundation for authentic assessment, student ownership of learning, and continuous peer support.

- Productive applications and cloud office tools

Within this group, we categorize the web office tools which allow the creation of text documents, visual presentations, databases, spreadsheets, calendars, etc., completely online without having to install anything on our devices and capable of being shared with others. Among the main possibilities these offer, we can highlight that we are no longer dependent on a computer, as the information is available online and the access to said tools is on the web. They have different uses: storage of documents, share and edit documents, organize and synchronize agendas, or publish information online.

The quintessential example of these tools is Google Drive, being as well one of the tools favored by university students for online collaboration [24]. Another prominent example for online file management and organization is Dropbox, by which we can create a basic, free account and save on the web (and share with whoever we decide) up to 2 GB worth of information, but this last tool does not allow the online edition of documents, only their storage, and sharing.

- Murals and Shared Whiteboards

Within this section, we include tools which permit users to add content in the same space/board to create and make contributions with information in a collaborative manner. They are usually very useful tools for the realization of schemes, brainstorming, and for the sharing of collaborative work processes. Some examples include Padlet or Mural.co.

Shared interactive boards become a privileged space for online evaluation, both continuous and final. The teaching staff can access easily to the contents submitted by students and assess all the process of shared work and construction. Peer-to-peer evaluation is also possible since the publications made by classmates can be accessed quickly and even comments can be made on them.

In this sense [55] found that the use of these tools in a collaborative way by students improves student confidence and motivation to write. This study demonstrated the benefits for teachers using shared murals. Teachers can quickly and easily assess students during the process and help them to improve continuously. In the same way, we find the research [71] with university students in which they state that evaluation through Padlet promotes creativity and collaborative learning in the classroom and optimizes student performance.

- Social Media

We refer to online environments where users publish and share digital objects around which a network of people is generated. The base of the collaboration lies on the shared object (videos, presentations, bookmarks, files, etc.), becoming an excellent medium for carrying out collaborative processes and the evaluation -both continuous and final- of these. Some examples are Flickr (image sharing), YouTube (video sharing), SlideShare (visual presentations), or Diigo (social bookmarking).

The teaching staff can access the creations of their students and evaluate the accomplished learnings. Students can in turn evaluate the contributions made by their classmates. Following the work of [18], we find some good examples about the use of YouTube in higher education for collaboration and online evaluation. The examples are: making students create a video as a part of the assessment, record a video and upload it on YouTube and use the comment section for discussion among students, have students search for videos related to questions posted at the end of lectures, showing students real-world examples and asking students to post video vignettes. On the other hand, with social bookmarking tools, such as Diigo, it is possible to review the information search process carried out, being this aspect one of the basic competencies that every university student should acquire.

- Social Networks

When we talk about social networking tools, we are specifically referring to those services or applications characterized by allowing the user to create a data profile about himself/herself online and share it with other users. The main potential of these tools for collaboration is the possibility of creating groups within them, which generates a "proper environment" for the communication and interaction between their members.

There is also the possibility of creating a specific social network in which to include the users participating and, thus, avoid participating on a general social network. This is the case of tools like Edmodo, where it is possible to collaborate and publish information, just as it is done in any other social network. One of the best-known examples of this type of tool is Facebook.

Within social networks and with some idiosyncratic characteristics, we find Twitter. Twitter is a telematic publishing tool, rather simple and based on the Web, which enables publishing small pieces of digital content. It is a useful tool for collaboration, since it allows us to communicate with others, be connected with a large number of users, foster social relationships, and share points of view or debate about certain topics of interest.

The possibilities for collaboration are endless, since with these tools, students can perform all types of communicative and collaborative processes, as different applications are integrated into one same space. Despite being a tool with great potential for collaboration, in previous studies we could observe [68] that students do not make extensive use of them for academic topics, leaving its use mainly for social matters.

The research on social networks and their possibilities in higher education shows that evaluation and, in particular peer evaluation, is a very interesting possibility among the educational uses that can be made of these tools. In research carried out by [85] with first-year university students from Taiwan University, peer assessment was implemented for the evaluation of learning to write in English. Among the most outstanding results of the research were that the incorporation of peer learning on Facebook is really engaging and effective at the university level. Besides, students improve their ability and knowledge to write in English thanks to the cooperative learning process generated on Facebook, and that the introduction of this tool in the assessment processes improves students' motivation and interest.

With regard to these topics, our research group finished a quantitative study in 2018 about the PLE and PLN of university students with 2054 answers from all the Spanish universities [24]. Our main research problem was the next: the type of digital tools and online strategies that our students use to communicate and collaborate with their peers in the academic work. The results show that in general, students prefer messaging apps (41%), followed by email (28%) and then social networking services (26%) for the academic tasks when they need to communicate with others. Less than 6% opted for video-conference tools. When students were asked about their preferences regarding the tools used to carry out group work or collaborative work, the majority of them use Google Drive (64.5%) followed by social networks (22.35%), virtual environments of their universities (7.98%); however, they did not even consider other tools like wikis or blogs. Moreover, students were asked about aspects they prioritize when working in teams. The majority of the students considered as priorities the fact of "building knowledge together" (58%), "interacting with others" (53%), and "sharing resources" (49%). In general, our conclusions showed that university students do not often use digital tools to collaborate as a learning strategy, but it is probably that teachers do not often use collaborative methods to teach in higher education.

3.4 Strategies of E-Assessment: Practical Experiences

The teaching–learning process, regardless of the level, must be minimally composed of 4 substantial elements: teaching activity (methodologies, strategies, and techniques), the didactic resources that both teachers and students will use, the evaluation to measure the scope of the objectives and students' learning, and finally, the feedback and report of the results of each student [5]. Therefore, the absence of one of the mentioned aspects will render the educational process incomplete and therefore, a good educational practice will not take place.

This simple proposal shows us the main elements which must be present in any educational modality, that is, regardless of whether students are educated in a virtual, face-to-face, or blended context. However, evaluation is one of the most widespread concerns which exist in institutions that offer digital training, due to different reasons such as phishing, storage and reuse of information, technical problems… among other possibilities [8, 84].

Specifically, the evaluation developed in a virtual learning context is called e-evaluation. It aims to offer a different vision than the results which are obtained with other types of more traditional evaluation in face-to-face contexts. Therefore, e-evaluation demands the use of a higher cognitive level, as well as some skills that would not be necessary to face in traditional evaluation methods [23]. Despite this idea, there are educational practices that turn classical evaluation formats, such as multiple-choice questions, into a type of digital evaluation [21, 66]. In other words, the support is changed, but not the strategy. Generally, e-evaluation is closely linked to higher education, since many universities or other training institutions offer their studies in double or triple modality (virtual, blended or face-to-face), being virtual modality one of the most frequently chosen types of learning in order to combine studies, work and family [12]. In the same sense, it is also affirmed that e-assessment facilitates the transfer of the academic to the working environment and, therefore, activates the learning in the present and throughout life.

At the same time, there is also the possibility of maintaining a digital evaluation in a face-to-face modality because it extracts more realistic results and presents new challenges to the agents involved, according to [36]. In addition, for this purpose, it is recommended using the "Bring Your Own Device" (BYOD) strategy. Specifically, this strategy consists of making a didactic use of the electronic devices (mobile phones, tablets or laptops) that each student owns. Likewise, BYOD makes assessment easy to continue outside the class, since it could be developed anywhere and anytime. In this way, it is guaranteed that an e-evaluation can be implemented, as a large room equipped with electronic devices will not be needed. Generally, student satisfaction with this type of assessment is usually high, since they positively value the commitment that exists between teachers and students to achieve quality training through constant feedback [3, 82].

E-evaluation has brought about different possibilities that distinguish it from any other type of evaluation. The one mentioned in the previous paragraph is one of them. However, others could be avoiding meltdown, tests on demand, students are able to progress at their learning pace, facilitating dialogue and participation among students, it favors the development of digital identity, improves motivation towards tests, etc. [51, 73].

The assessment has to be in harmony with the methodology followed during the teaching–learning process and one of the biggest challenges that e-assessment presents is when establishing strategies that evaluate collaborative learning in an objective and real way. This collaboration represents a specific model of e-learning, so the evaluation must be coherent with the instructional design. The complexity of this fact is due to two elements: learning arises through group interaction and in virtual environments. For this, students must participate actively in different assessment processes, in order to have more sources of evaluation and obtain authentic results [78]. Mainly, the recommended strategies for e-assessment in a collaborative environment combine both formative and summative evaluations, inasmuch it must focus on both processes and results [28, 83]. In addition, it is also pertinent to develop a self-assessment and a peer or co-assessment process [28, 29].

3.4.1 E-Formative Assessment

Firstly, formative evaluation is the one that extends throughout the teaching–learning process of certain contents. From it, immediate, constant and quality feedback will be offered among the agents involved in the educational task, mainly students and teachers [6]. Formative evaluation is one of the main guides for the teacher because he will be able to modify or maintain the teaching plan, making decisions based on the real results of his actions. This type of information can be useful to inform us about the progress of students but in the same way about the collaboration itself (the interaction, the development of the tasks, the responsibility of students, the adjustment of their roles, etc. all the dimensions of collaborative learning).

Feedback is the key aspect of this e-assessment strategy, as students will be able to complete the corresponding surveys or tasks online and obtain an automatic response. Likewise, these grades will also be available to teachers, which will also help them determine the learning and progress of each student [83] and the progress of collaborative groups.

It should be noted that despite the type of electronic formative evaluation that is carried out, the commitment and satisfaction of the students are high. In [3] it is evident that despite implementing three different options (online knowledge survey; online questions and answers generated by students; and reflective electronic journals), students were always positively valued, so a varied use of them is recommended.

In addition to this, other options for conducting formative e-evaluations are through gamification techniques, for example with the use of digital badges [27], leveling up the ladder of challenges [7], etc. Another possibility is through learning analytics, which informs about the student's progress in real-time, so they are also a valid option to favor the development of formative e-evaluation [49].

Generally, some of the most widely used tools to develop a formative e-assessment process are social networks. Their defining instantaneous character makes them a valuable option for asynchronous feedback on student progress. There are numerous practices based on popularly known social networks, for example, through an exchange of comments on *Facebook* to guide the practice of different workgroups [65] or the use of Twitter and its function as a tool to evaluate formatively has also generated significant benefits in relation to academic performance [81], among other options with a greater bonding to the educational field, such as conducting online questionnaires through Edmodo [63].

Despite the extensive use of social networks to implement formative e-assessment, it should be noted that there is also a large bibliography that relates this type of assessment to the use of "collaborative environments" [33, 79], in which there are tools to generate automatic feedback [72]. In addition, some proposals based on "web office tools" such as *Google Drive* have also been found, mainly due to their synchronous character [87] or digital forums such as the ones we can find in the LMS, so long as professors can make the pertinent recommendations to make the task a collaborative one [39]. Finally, we should mention the use of "videoconferencing tools", in order to promote the interaction with tutors and professors, being as well a useful tool for students to receive feedback. It should be noted that sometimes the videoconference is developed around a semi-structured interview [58].

3.4.2 E-Summative Assessment

On the other hand, summative evaluation is defined as the results obtained at the end of a lesson, course or project, to determine the degree of learning achievement and, therefore, the scope of predetermined objectives [83]. The summative e-evaluation strategy usually employs instruments similar to those used in a face-to-face modality, that is, a final standard exam or a test, but on e-learning processes it is also frequent the delivery of a final report or task [48, 77].

In regard to the use of online questionnaires as an instrument for the e-summative assessment, there is the possibility of using a specific digital tool of a LMS [21, 74]. However, there are other alternatives such as Google forms [53] or other digital tools for this purpose. However, one of the risks that are most associated with this type of instrument is the lack of a reliable e-assessment system [90].

On the other hand, [44] recommend that not only questionnaires should be used to perform the summative e-evaluation, but other alternatives must be offered. Thus, [52] expose that students of the degree in Physics can do a voluntary exercise of a simulation, whose score will be added directly to the summative grade. To do this, students must previously be trained in the use of software, the realization of online tutorials, among other tasks.

This type of summative e-assessment can also be developed with synchronous tools like videoconference, so we can design oral exams or presentations where the student must show some results, present his/her task, or explain his learnings, while the teacher can ask some questions in real-time. We have the possibility to record the interaction by videoconference as evidence of the exam, but also as an example to other students building a community of resources which includes students' presentations [96, 97]

In relation to these two evaluation strategies led by teachers, formative and summative, in [17] it is emphasized that a virtual learning context must use both, since dialogue, learning structuring, understanding of progress, among other reasons, are favored. And both can be part of the final mark. In addition to the summative and formative e-evaluation, the two strategies that must also be used in a collaborative online learning context are self-evaluation and peer evaluation, which favor an even more active role for students [75].

3.4.3 E-Self Assessment

Firstly, self-assessment is considered as an interesting strategy so that students can compare the evaluations provided by teachers, although it is also recommended that teachers evaluate their own work periodically [61]. This e-assessment strategy involves a reflexive process from which the subject must identify the strengths and weaknesses in relation to accomplishing the objectives [56]. For this, five levels of reflection are recommended: to inform yourself, respond, relate, reason, and reconstruct, which can be put into practice in combination with the use of different digital tools [2]. In addition, this strategy used in virtual contexts not only implies improving knowledge about one's own progress, but also increases interest and involvement in the subject [50].

E-self-assessment is a possibility that, as reflected, can significantly improve the motivation towards learning and the promotion of critical thinking. Conversely, the opposite consequence can also occur, that is, students become demoralized and lose interest in the subject [26]. Therefore, it should be taken into account that self-assessment is not decisive, but a formative evaluation strategy carried out by students to maintain or change their study or work methods.

Generally, higher education is a level where there are various digital self-assessment practices. For example, in Law degrees [14] or Education [20].

Digital self-assessment, like the rest of e-evaluation strategies, can be developed both in a face-to-face context with digital resources or in a virtual context using the online platform with the corresponding complements. In relation to the resources used to implement digital self-assessment, different options have to be highlighted. Collaborative e-portfolios—though a given "web office tool"—are a widely used alternative, since the members of a team can reflect on their autonomous and group work. In this way, the group has the possibility to reflect criticisms or suggestions and thus, improve their subsequent work, whether individual or collective [43, 57]. As [22] conclude, the portfolio, along with social media are interesting alternatives to traditional assessment and they are confident that these tools are useful in collaborative processes.

On the other hand, there are also different digital applications that can be downloaded on different digital devices, and thus carry out digital self-assessment. Some examples of this case are: the use of the "seesaw" platform, an open educational resource that consists of an e-portfolio that also values family vision since this educational agent can observe and provide feedback on the work done [11]; the use of the "Socrative" App for digital self-evaluation of both students and teachers. permits obtaining better assessment by students due to the increase in academic performance and the satisfaction generated [13]; implementation of interactive manuals and online questionnaires through the Moodle learning management system (LMS) [9], among others. Finally, we can use a digital rubric or a simple test; it is very easy from a technical point of view with the tools that we have in our LMS. However, the difficulties are to concrete the educational use of this information and especially the veracity of the information in case it is shared with teachers without anonymity. In other words, the students will be honest if the results are only seen by them, for their own personal use, but they will probably not be totally sincere if teachers read the answers and these are part of the final mark. Self-assessment can be very useful, understood as a way to improve the own working processes (for both teachers and students, teaching and learning activities) and to make better decisions in the future.

3.4.4 E-Peer Assessment

Electronic peer assessment is defined as the process through which students can analyze and assess the process and the results of a partner or group of the same level [75]. The cited work also exposes that there are conceptual errors in relation to certain evaluation strategies like peer-assessment and co-assessment (or evaluation). Co-evaluation differs from peer-evaluation in that the first consists of the group assessment of students together with teachers on the scope of the established objectives, and the second only involves the analysis and assessment between two students or between groups of them. Likewise, in an erroneous way, [10] point out that co-evaluation is understood as a peer-evaluation of the work that other colleagues have done. In the end, in spite of different definitions, we can agree that peer-evaluation is a strategy that provides students with greater autonomy and prominence to contrast, reflect and complete their training on a subject [76] and therefore, as [28] point out, it is a key strategy in the e-assessment of a collaborative work.

Over time, different proposals have been developed to contribute to the effective development of online or blended training through digital peer evaluation [41, 94]. As it has been proven, the tools and methodologies used in this e-evaluation strategy are diverse. Some examples are the use of blogs [42]; the use of e-co-rubrics [30, 45]; Web-based peer-tutoring system called "Opal" by [19]; practices with video like web-based video-annotation by [37], or online video sharing [31] or even through synchronous videos using "videoconferencing tools" [64]; using a private group on a "social network" like Facebook [15, 40]; and finally, some "collaborative environments" such as Moodle have optimal tools to develop peer-assessment [1].

Despite the benefits that the above-mentioned practices have produced in their contexts, it should also be mentioned that online peer tutoring can be a strategy that generates conflicts and behavioral problems among the group members [95]. However, as previously noted, it is a very useful strategy to reflect, discuss, and solve intra/inter-group problems that may exist. In relation to this e-assessment strategy, it is necessary to emphasize that MOOCs (Massive Open Online Courses), specifically, in its cMOOC modality, evaluation always favors interaction from a formative perspective. Therefore, this virtual learning modality has digital peer evaluation as its default strategy to determine progress [80].

3.5 Arriving to Practical Recommendations: Design and Implement E-Assessment in Online Collaboration

Throughout the text, different collaborative techniques for working in a virtual modality have been explained, several tools that can be used for digital assessment, some examples of collaborative environments and the four types of e-assessment have been studied in-depth and many examples based on evidence of each one have been provided.

3.5.1 Discussion

In general, one of the types of e-assessment that presented more difficulties when determining the most widely used tools and strategies was summative e-assessment. In the text, it is mentioned that this type of e-assessment could be summarized in the final reports of a task or the in digital questionnaires used [48, 77], however, there are works which state that digital questionnaires are optimal for e-formative assessment, not for e-summative assessment [86]. In the light of this, one of the prospective lines of research that derives from this work appears, that is, summative e-evaluation in collaborative work environments based on real practices. Along the same lines, e-self-assessment is also one of the types of e-assessment that should be delved into, since it is a crucial process to complement the rest of evaluation strategies and, above all, to self-analyze group practice and thus, redefine roles, redistribute work and, in short, learn from your own practice from a self-critical perspective.

One of the most striking results is that "wikis", "blogs" and "social networks" are only linked with two e-assessment types. For example, "blogs" could be justified by the students' refusal to use them, since, as it has been exposed in the text, the students of the Spanish universities do not value "wikis" and "blogs" as preferential when collaborating online [71]. However, students may not value these options due to lack of knowledge, since their professors may not have used them. Therefore, it is important to select the most suitable e-tool according to the objectives and strategies.

On the contrary, if the students' preferences for collaborating online are considered, they highlighted Google Drive (64.5%), messaging tools (41.19%), and media tools (25.85%) [24]. In this case, the most widely used tool by students coincides with the most used in the different types of e-assessment, that is, "web office tools".

Despite the fact that students point out that messaging tools are their second preferred option for collaborating online, they do not usually use official channels such as the virtual environments of their universities, since only 7.98% consider them as a preference. This fact shows that students prefer to use other private communication channels such as social networks, which are also linked to the two types of e-assessment- in which interaction and feedback abound- that is, e-formative assessment and e-peer assessment.

A striking aspect of the virtual environments we have mentioned is that, while all of them have proven useful for the four types of e-assessment addressed in these pages, they are, notwithstanding their versality, not among the tools favored by students, as shown in the previous paragraphs. Instead, a collaborative space such as an LMS of a university usually has a wide range of digital tools that facilitate joint construction, interaction with others, and sharing resources, that is, the highest priority objectives according to students when it comes to teamwork [71].

3.5.2 Conclusions

After this discussion of results, it is considered of interest to present a figure of virtual learning that reflects the four e-evaluation strategies mentioned throughout these pages, that is, formative, summative, self-evaluation, and between peers. We have experimented with all of them in our online master's degree in educational technology. We use a virtual LMS (Moodle) to study the subjects and the "Adobe Connect" software for synchronous communication (exhibitions, debates between students, etc.). In addition, each subject can offer specific digital tools to develop and evaluate certain activities because our instructional design is based on the combination of both individual and group work; and the latter case, the groups can be traditional or collaborative ones in relation to different tasks.

To promote collaboration, we use different tools like social networks or instant messaging; documents on the cloud to promote collaborative edition; collaborative virtual spaces like BSCW or Google Drive to share information; discussions in real-time using videoconferences or delayed, using forums. The next Fig. 3.3 is like an abstract of all the concepts that we have explained in this chapter, considering that all these possibilities of e-assessment will be implemented by different tools on our LMS or other external tools, which we have used as examples in our analysis.

Therefore, it is considered of interest to cross the data referring to the mentioned tools with the four types of evaluation according to the different practices exposed in the work. In this way, significant conclusions can be drawn regarding the most used tools in accordance with the type of e-assessment. Specifically, the instrument used is a two-way table, since it is an optimal instrument to reduce and express the existing data and therefore, achieve the stated objective. It should be noted that the results presented are directly based on the bibliographic review carried out and the references used throughout the work. The seven digital tools that have been exposed in the present work have been crossed with the four types of e-evaluation

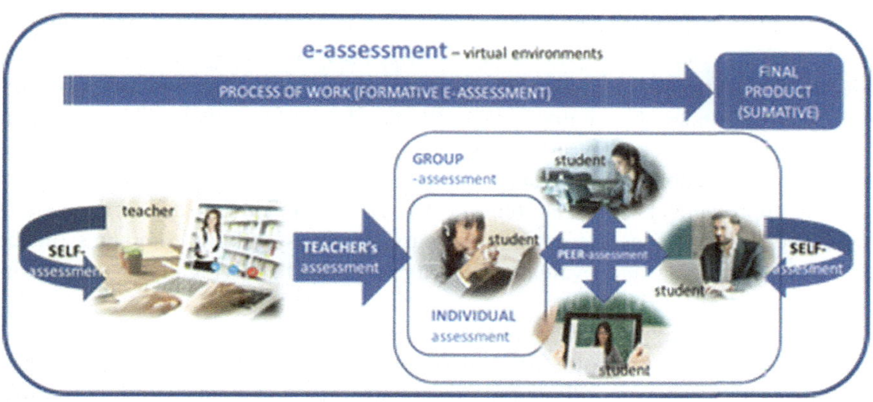

Fig. 3.3 E-assessment in collaborative group

Table 3.3 E-tools for e-assessment

	E-formative assessment	E-summative assessment	E-self assessment	E-peer assessment
LMS	X	X	X	X
Collaborative work-spaces	X	X	X	X
Web 2.0 tools				
– Wikis	X			X
– Blogs	X		X	X
– Video-conference		X	X	X
– Cloud office tools	X	X	X	X
– Social media				X
– Social networks	X			X

also mentioned. A total of 15 crosses have been extracted; these are shown in Table 3.3.

Firstly, we must justify that the crossings that have been made between the types of e-assessment and the digital tools have been in accordance with the existing bibliography and experiences, some of them cited in this chapter. As the table shows, a tool does not only have to be linked to a type of evaluation, since depending on the use made of it, different objectives can be achieved. For example, the use of an e-portfolio through a "web office tool" could be a summative evaluation instrument if it is delivered as the result of a task; or a formative assessment tool to observe student progress and provide feedback; among other possibilities for peer assessment or self-assessment.

In relation to the collaborative tools, LMS, and web office tools, we find that due to their characteristics and their possibilities they allow the four types of evaluation mentioned above.

Wikis allow e-formative assessment because it is possible to follow and comment on the tasks carried out by each student. Also, peer-to-peer assessment is possible because wikis allow us to revise and modify what has been suggested and published by other classmates. Not unlike wikis, Blogs allow e-formative and peer-to-peer assessment; moreover, they allow e-self assessment. On a blog, it is possible to provide quick feedback to the students and they offer the opportunity for the student to make improvements on the contents posted. Also, being able to access the content published by other classmates and by yourself allows making comments on the posts and reviewing and evaluating among peers and individually the learnings accomplished.

Videoconference simulates a real communicative process, and, for this reason, this tool allows the four types of e-assessment. In order to do e-self assessment, video-conference allows the recording of the communication and thanks to this recording, students are able to improve after reviewing their performances.

Regarding social media, and as one of its main purposes is file sharing (image, video, links, presentations), peer-to-peer assessment is a great option through these shared elements.

To conclude the explanation of this table, social networks are an accurate way for both e-formative and peer-to-peer assessment. These tools allow for continuous monitoring and feedback of student work by the teacher and by the students themselves. This enriches and improves the training processes and the evaluation carried out.

The most useful and flexible tool is the collaborative environment or a LMS, as since we can develop different types of e-assessment without changing the digital tool, we have all the possibilities to implement our collaborative instructional design. The use of tools on the cloud to promote the edition in groups and the process of sharing information is bound to be highly interesting.

Finally, we consider that this approach to e-assessment can open future lines of research because this final proposal is the result of our experience and previous research explained in the chapter, but it must be validated in different university contexts.

References

1. Amendola, D., & Miceli, C. (2016). Online physics laboratory for university courses. *Journal of E-Learning and Knowledge Society, 12*(3), 75–85. Retrieved from: https://bit.ly/2TmuD5g.
2. Amhag, L. (2020). Student reflections and self-assessment in vocational training supported by a mobile learning hub. *International Journal of Mobile and Blended Learning, 12*(1), 1–16. https://doi.org/10.4018/ijmbl.2020010101.
3. Bahati, B., Fors, U., Hansen, P., Nouri, J., & Mukama, E. (2019). Measuring learner satisfaction with formative e-assessment strategies. *International Journal of Emerging Technologies in Learning, 14*(7), 61–79. https://doi.org/10.3991/ijet.v14i07.9120.
4. Barber, W., King, S., & Buchanan, S. (2015). Problem based learning and authentic assessment in digital pedagogy: Embracing the role of collaborative communities. *Electronic Journal of E-Learning, 2*(13), 59–67. Retrieved from https://bit.ly/2TT7CHB.
5. Barbosa, H., & García-Peñalvo, F. J. (2005). Importance of online assessment in the e-learning process. In *6th International Conference on Information Technology Based Higher Education and Training* (pp. F3B1–F3B6). IEEE. https://doi.org/10.1109/ithet.2005.1560287.
6. Basson, S. N., Van der Watt, H. C., & Hancke, C. H. (2014). E-assessment as tool to improve students' learning experience and pass rates in mechanical engineering. In L. Gómez, A. López, & I. Candel (Eds.), *EDULEARN14 Proceedings* (pp. 5237–5245). Valencia: IATED Academy. Retrieved from https://bit.ly/2RosAMb.
7. Bezzina, S. (2019). Games, design and assessment: how game designers are doing it right. In L. Elbaek, G. Majgaard, A. Valente, & S. Khalid (Eds.), *Proceedings of the European Conference on Games-Based Learning* (pp. 67–73). Frankfurt: Dechema e.V. Retrieved from https://bit.ly/30r3fp8.
8. Cabero, J. (2017). *La evaluación en la era digital.* Madrid: Síntesis. Retrieved from https://bit.ly/2Ng8eTY.
9. Candelas-Herías, F. A., Gil, P., Jara, C. A., Corrales, J. A., & Baquero, M. A. (2011). Recursos digitales interactivos para la asignatura de sistemas de trans-porte de datos para potenciar el aprendizaje autónomo y la autoevaluación. In J. D. Álvarez, M. T. Tortosa, & N. Pellín

(Eds.), *Redes de investigación docente universitaria* (pp. 2505–2533). Universidad de Alicante. Retrieved from https://bit.ly/2QVMaAD.

10. Castillo, S., & Cabrerizo, J. (2003). *Evaluación educativa y promoción escolar*. España: Pearson Educación.
11. Chaljub, J. M. (2019). La plataforma digital Seesaw. Su integración en una clase dinámica. *Pixel-Bit: Revista de Medios y Educación, 54*, 107–124. https://doi.org/10.12795/pixelbit.2019.i54.06.
12. Chandrasekaran, S., Badwal, P., Thirunavukkarasu, G., & Littlefair, G. (2016). Collaborative learning experience of students in distance education (conference paper). In: *8th International Symposium on Project Approaches in Engineering Education*. Portugal: Guimarães. Retrieved from https://bit.ly/3agmtCa.
13. Cosi, S., & Voltas, N. (2019). Evaluación formativa en estudiantes universitarios mediante tecnologías digitales: El rol del alumno en su propio proceso de enseñanza-aprendizaje. In R. Roig-Vila (Ed.), *Investigación e innovación en la enseñanza superior* (pp. 113–123). Barcelona: Octaedro. Retrieved from https://bit.ly/2Tpfktj.
14. De Barrón, P. (2019). La autoevaluación en línea en el estudio del Derecho. In A. M. Delgado & I. B. De Heredia (Eds.), *La docencia del Derecho en la sociedad digital* (pp. 231–234). Universitat Oberta de Catalunya. Retrieved from https://bit.ly/3ackizG.
15. Demir, M. (2018). Using online peer assessment in an instructional technology and material design course though social media. *Higher Education, 75*(3), 399–414. https://doi.org/10.1007/s10734-017-0146-9.
16. DiNucci, D. (1999). *Fragmented Future. Design & New Media*. Darcyd. Retrieved from https://bit.ly/2tFMxWS.
17. Dorrego, E. (2006). Educación a distancia y evaluación del aprendizaje. *RED: Revista de Educación a Distancia, 6*. Retrieved from https://bit.ly/2QQQp04.
18. Duffy, P. (2008). Engaging the YouTube google-eyed generation: Strategies for using Web 2.0 in teaching and learning. *The Electronic Journal of E-Learning, 6*(2), 119–130. Retrieved from: https://bit.ly/3efHUEI.
19. Evans, M. J., & Moore, J. S. (2013). Peer tutoring with the aid of the Internet. *British Journal of Educational Technology, 44*(1), 144–155. https://doi.org/10.1111/j.1467-8535.2011.01280.x.
20. Flores-Lueg, C., & Roig-Vila, R. (2016). Diseño y validación de una escala de autoevaluación de competencias digitales para estudiantes de pedagogía. *Pixel-Bit: Revista de Medios y Educación, 48*, 209–224. Retrieved from https://bit.ly/36USDRV.
21. Gamage, S. H. P. W., Ayres, J. R., Behrend, M. B., & Smith E. J. (2019). Optimising moodle quizzes for online assessments. *International Journal of STEM Education, 6*(1). https://doi.org/10.1186/s40594-019-0181-4.
22. García-Chitiva, M. P., Suárez-Guerrero, C. (2019). Estado de la investigación sobre colaboración en Entornos Virtuales de Aprendizaje. *Píxel-Bit Revista de Medios y Educación, 56*, 169–192. https://doi.org/10.12795/pixelbit.2019.i56.09.
23. Grisp, G., Guàrdia, L., & Hillier, M. (2016). Usisng e-Assessment to enhance student learning and evidence learning outcomes. *International Journal of Educational Technology in Higher Education, 13*(1). https://doi.org/10.1186/s41239-016-0020-3.
24. Gutiérrez, I., Román, M., & Sánchez, M. (2018). Estrategias para la comunicación y el trabajo colaborativo en red de los estudiantes universitarios. *Comunicar, 54*, 91–100. https://doi.org/10.3916/C54-2018-09.
25. Häkkinen, P. (2002). Challenges for design of computer-based learning environments. *British Journal of Educational Technology, 33*(4), 461–469. Retrieved from: https://bit.ly/3d7JMiI.
26. Han, C., & Fan, Q. (2020). Using self-assessment as a formative assessment toolibarra in an English-Chinese interpreting course: Student views and perceptions of its utility. *Perspectives: Studies in Translation Theory and Practice, 28*(1), 109–125. https://doi.org/10.1080/0907676x.2019.1615516.
27. Hennah, N., & Seery, M. (2017). Using digital badges for developing high school chemistry laboratory skills. *Journal of Chemical Education, 94*(7), 844–848. https://doi.org/10.1021/acs.jchemed.7b00175.

28. Hernández, N., Muñoz, P. C., & González, M. (2018). La e-evaluación en el trabajo colaborativo en entornos virtuales: Análisis de la percepción de los estudiantes. *Edutec, 65*, 16–28. https://doi.org/10.21556/edutec.2018.65.997.
29. Hernández, J. S., Tobón, S., Ortega, M. F., & Ramírez, A. M. (2018). Evaluación socioformativa en procesos de formación en línea mediante proyectos formativos. *Educar, 54*(1), 147–163. https://doi.org/10.5565/rev/educar.766.
30. Hoffman, B. (2019). The influence of peer assessment training on assessment knowledge and reflective writing skill. *Journal of Applied Research in Higher Education, 11*(4), 863–875. https://doi.org/10.1108/jarhe-01-2019-0004.
31. Hwang, G. H., Chen, B., & Sung, C. W. (2019). Impacts of flipped classrooms with peer assessment on students effectiveness of playing musical instruments—Taking amateur erhu learners as an example. *Interactive Learning Environ-Ments, 27*(8), 1047–1061. https://doi.org/10.1080/10494820.2018.1481105.
32. Joy, E., Evelyn, J., & Myers, B. (2009). Using wikis and blogs for assessment in first-year engineering. *Campus-Wide Information Systems, 26*(5), 424–432. https://doi.org/10.1108/10650740911004831.
33. Kartal, E. E., Dogan, N., Irez, S., Cakmakci, G., & Yalaki, Y. (2019). A five-level design for evaluating professional development programs: Teaching and learning about nature of science. *Issues in Educational Research, 29*(2), 402–426. Retrieved from https://bit.ly/3cCmApx.
34. Kauppi, S., Muukkonen, H., Suorsa, T., & Takala, M. (2020). I still miss human contact, but this is more flexible—Paradoxes in virtual learning interaction and multidisciplinary collaboration. *British Journal of Educational Technology, Early View*. https://doi.org/10.1111/bjet.12929.
35. Kelly, D., Baxter, J. S., & Anderson, A. (2010). Engaging first-year students through online collaborative assessments. *Journal of Computer Assisted Learning, 26*(6), 535–548. https://doi.org/10.1111/j.1365-2729.2010.00361.x.
36. Küppers, B., & Schroeder, U. (2016). Bring your own device for e-assessment—A review. In L. Gómez, A. López & I. Candel (Eds.) *EDULEARN16 Proceedings* (pp. 8770–8776). Valencia: IATED Academy. https://doi.org/10.21125/edulearn.2016.0919.
37. Lai, C. Y., Chen, L. J., Yen, Y. C., & Lin, K. Y. (2020). Impact of video annotation on undergraduate nursing students' communication performance and commenting behavior during an online peer-assessment activity. *Australasian Journal of Educational Technology, 36*(2), 71–88. https://doi.org/10.14742/ajet.4341.
38. Lee, H. (2008). *Students' perceptions of peer and self assessment in a higher education online collaborative learning environment* (Thesis Dissertation, The University of Texas at Austin). University of Texas Libraries. Retrieved from https://bit.ly/3d9o4uU.
39. Lezcano, L., & Vilanova, G. (2017). Instrumentos de evaluación de aprendizaje en entornos virtuales: perspectiva de estudiantes y aportes de docentes. *Informe Científico Técnico UNPA, 9*(1), 1–36. https://doi.org/10.22305/ict-unpa.v9i1.235.
40. Lin, G. Y. (2018). Anonymous versus identified peer assessment via a Facebook-based learning application: Effects on quality of peer feedback, perceived learning, perceived fairness, and attitude toward the system. *Computers and Education, 116*, 81–92. https://doi.org/10.1016/j.compedu.2017.08.010.
41. Lin, Y., & Lin, Y. (2020). Optimal design of online peer assessment system. In J. S. Pan, J. Li, P. W. Tsai, & L. C. Jain (Eds.), *Advances in intelligent information hiding and multimedia signal processing* (pp. 217–224). Singapore: Springer. https://doi.org/10.1007/978-981-13-9710-3_23.
42. Lizandra, J., & Suárez, C. (2017). Trabajo entre pares en la curación digital de contenidos curriculares. *RELATEC, 16*(2), 177–191. Retrieved from https://bit.ly/38avaMM.
43. López-Meneses, E., Vázquez, E., & Jaén, A. (2017). Los portafolios digitales grupales: Un estudio diacrónico en la Universidad Pablo Olavide (2009–2015). *Revista De Humanidades, 31*, 123–152. https://doi.org/10.5944/rdh.31.2017.19076.
44. Malach, J., & Švrčinová, V. (2018). Theoretical and methodological basis of assessment of pedagogical digital competences. In A. Andreatos, C. Sgouropoulou, & K. Ntalianis (Eds.), *Proceedings of the European Conference on E-Learning* (pp. 354–360). England: Academic Conferences Limited. Retrieved from https://bit.ly/38cLnko.

45. Martínez, D. D., Cebrián, D., & Cebrián, M. (2016). Assessment of teaching skills with e-Rubrics in Master of teacher training. *JETT, 7*, 120–141. Retrieved from https://bit.ly/36WELGJ.
46. Martínez, F. & Prendes, M. P. (2008). Estrategias y espacios virtuales de colaboración para la enseñanza superior. *Revista Internacional de Ciencias Sociales y Humanidades SOCIOTAM, 23*(2), 59–90. Retrieved from https://bit.ly/2znJNAC.
47. McLeod, S., & Lehmam, C. (2012). *What school leaders need to know about digital technologies and social media*. San Francisco: Jossey-Bass.
48. Mellado, M. E. (2007). Portafolio en línea: Una herramienta de desarrollo y evaluación de competencias en la formación docente. *Educar, 40*, 69–89. Retrieved from https://bit.ly/2AD06JR.
49. Menchaca, I., Guenaga, M., & Solabarrieta, J. (2018). Learning analytics for formative assessment in engineering education. *The International Journal of Engineering Education, 34*(3), 953–967. Retrieved from https://bit.ly/2RoyiOg.
50. Míguez, M. I., & Dafonte, A. (2018). Las aplicaciones de cuestionarios de autoevaluación en el aula a través de dispositivos móviles. In J. Valverde (Ed.), *Campus digitales en la educación superior* (pp. 245–254). Retrieved from https://bit.ly/2FNqlgb.
51. Mimirinis, M. (2019). Qualitative differences in academics conceptions of e-assessment. *Assessment and Evaluation in Higher Education, 44*(2), 233–248. https://doi.org/10.1080/02602938.2018.1493087.
52. Montoya, M. M., & Rubio, M. A. (2018). Prácticas de simulación en la asignatura "teoría de circuitos y electrónica". In M. C. Ortega, M. A. López, P. Amor (Eds.), Innovación educativa en la era digital (pp. 177–179). Madrid: UNED. Retrieved from https://bit.ly/30tsTcN.
53. Mora, F. (2011). Experiencia en el uso de encuestas en línea para la evaluación diagnóstica y final de un curso virtual. *Tecnología en Marcha, 24*(4), 96–104. Retrieved from https://bit.ly/36fOGYa.
54. Morales Chan, M., Barchino Plata, R., Medina, J. A., Alario-Hoyos, C., & Hernandez Rizzardini, R. (2019). Modeling educational usage of cloud-based tools in virtual learning environments. *IEEE Access, 7*, 13347–13354. https://doi.org/10.1109/access.2018.2889601.
55. Moreira, H. (2019). *Implementing the writing process through the collaborative use of Padlet* (Master Thesis, University Casa Grande, Guayaquil). Repositorio digital Universidad Casa Grande. Retrieved from https://bit.ly/2TCrqyF.
56. Mosqueda, D. L. (2018). Uso de cuestionarios en Moodle para la autoevaluación de los conocimientos matemáticos. In J. Valverde (Ed.), *Campus digitales en la educación superior* (pp. 679–684). Retrieved from https://bit.ly/2FNqlgb.
57. Navarro, J. (2016). Experiencia el porafolio de mi clase: un mix de aula inversa, portafolios digitales y autoevaluación. In R. Roig-Vila (Ed.), *Tecnología, innovación e investigación en los procesos de enseñanza-aprendizaje* (pp. 2810–2814). Barcelona: Octaedro. Retrieved from https://bit.ly/2NssxOu.
58. O'Donovan, J., Maruthappu, M., & Edwards, J. (2015). Distant peer-tutoring of clinical skills, using tablets with instructional videos and Skype: A pilot study in the UK and Malaysia. *Medical Teacher, 37*(5), 463–469. https://doi.org/10.1016/j.aogh.2015.02.949.
59. O'Reilly, T. (2005, September 30). What is Web 2.0: Design patterns and business models for the next generation of software. Mediaedu.typepad. Retrieved from https://bit.ly/2M0zuFi.
60. O'Reilly, T., & Battelle, J. (2009). Web squared: Web 2.0 five years on [Report]. In *Web2.0 summit*. Retrieved from https://bit.ly/2NNGIh3.
61. Ochoa, L., & Moya, C. (2019). La evaluación docente universitaria: Retos y po-sibilidades. *Folios: Revista de la Facultad de Humanidades, 49*, 41–60. Retrieved from https://bit.ly/3a80MUR.
62. Olofsson, A. D., Ola Lindberg, J., & Eiliv Hauge, T. (2011). Blogs and the design of reflective peer-to-peer technology-enhanced learning and formative assessment. *Campus-Wide Information Systems, 28*(3), 183–194. https://doi.org/10.1108/10650741111145715.
63. Oversby, J., Sanders, J. Talib, C. A., Thoe, N. K., & Esa, N. (2019). Question generating supported by blended learning platform: Issues of social justice for environmental education.

Eurasia Journal of Mathematics, Science and Technology Education, 15(5), em1709. https://doi.org/10.29333/ejmste/105848.
64. Page, R., Hynes, F., & Reed, J. (2019). Distance is not a barrier: The use of videoconferencing to develop a community of practice. *Journal of Mental Health Training, Education and Practice, 14*(1), 12–19. https://doi.org/10.1108/jmhtep-10-2016-0052.
65. Piantola, M. A. F., Moreno, A. C. R., Matielo, H. A., Taschner, N. P., Cavalcante, R. C. M., Khan, S., et al. (2018). Adopt a bacterium—An active and collaborative learning experience in microbiology based on social media. *Brazilian Journal of Microbiology, 49*(4), 942–948. https://doi.org/10.1016/j.bjm.2018.04.005.
66. Pramadya, W., Riyadi, R., & Indriati, D. (2019). Self-assessment profile on statistics using computer-based mathematical summative test. *Journal of Physics, 1188*(1). https://doi.org/10.1088/1742-6596/1188/1/012053.
67. Prendes-Espinosa, M. P. (2003). Aprendemos…¿cooperando o colaborando? Las claves del método. In F. Martínez (Ed.), *Redes de comunicación en la enseñanza: Las nuevas perspectivas del trabajo cooperativo* (pp. 93–127). Barcelona: Paidós.
68. Prendes-Espinosa, M. P., Gutiérrez-Porlán, I., & Castañeda-Quintero, L. (2015). Perfiles de uso de redes sociales: Estudio descriptivo con alumnado de la Universidad de Murcia. *Revista Complutense de Educación, 26*. https://doi.org/10.5209/rev_rced.2015.v26.46439.
69. Prendes-Espinosa, M. P., Castañeda-Quintero, L., Gutiérrez-Porlán, I. & Sánchez-Vera, M. M. (2017). Personal learning environments in future professionals: Nor natives or residents, just survivors. *International Journal of Information and Education Technology, 7*(3), 172–179. https://bit.ly/2P82Zq3.
70. Project CARMA. (2018). CARMA Toolkit (RMA and other non-formal learning techniques). A step-by-step guide for implementing collaborative learning to increase student motivation and participation. Erasmus + Programme of the European Union. Retrieved from https://bit.ly/3ek3cRy.
71. Reka, C., & Mahmud, M. M. (2018). Padlet: A technology tool for the 21st century students skills assessment. In M. A. Safari, M. A. Bagus, & M. S. Ence (Eds.), In: *Proceeding Book of 1st International Conference on Educational Assessment and Policy* (Vol. 1, pp. 101–107). Center for Educational Assessment (Puspendik). Retrieved from https://bit.ly/2LZJ48Q.
72. Remesal, A., Colomina, R., Mauri, T., & Rochera, M. J. (2017). Uso de cuestionarios online con feedback automático para la e-innovación en el alumnado universitario. *Comunicar, 51*, 51–60. https://doi.org/10.3916/c51-2017-05.
73. Ridgway, J., McCusker, S., & Pead, D. (2004). *Literature review of e-assessment* (Report n.° 10). Bristol: Futurelab. Retrieved from https://bit.ly/2NwSxbn.
74. Rodas, P. (2015). Aprendizaje y evaluación de conocimientos prácticos a través de cuestionarios Moodle. In M. Villca & A. Carreras (Eds.), *Docencia virtual y experiencias de innovación docente: Entornos b-learning y e-learning* (pp. 216–226). Barcelona: Hygens D.L.
75. Rodríguez, G., Ibarra, M. S., & García, E. (2013). Autoevaluación, evaluación entre iguales y coevaluación: conceptualización y práctica en las universidades españolas. *Revista de Investigación en Educación, 11*(2), 198–210. Retrieved from https://bit.ly/30nG5zP.
76. Rodríguez-Migueles, A., & Hernández-Yulcerán, A. (2014). Desmitificando algunos sesgos de la autoevaluación y coevaluación en los aprendizajes del alumnado. *REXE: Revista de Estudios y Experiencias en Educación, 13*(25), 13–31. Retrieved from https://bit.ly/35X9sdk.
77. Rojas, F. R., Valero, V., & Cortés, G. E. (2014). Evaluación virtual: Examen de ubicación de inglés en línea de la UAM-A. *Relingüística Aplicada, 14*. Retrieved from https://bit.ly/3bNhFDV.
78. Romeu, T., Romero, M., & Guitert, M. (2016). E-assessment process: Giving a voice to online learners. *International Journal of Educational Technology in Higher Education, 13*(20), 1–14. https://doi.org/10.1186/s41239-016-0019-9.
79. Sacristán, M., Garrido-Vega, P., Alfalla-Luque, R. & González-Zamora, M. M. (2010). De la evaluación sumativa a la formativa a través de las plataformas de enseñanza virtual. In R. Del Pozo (Ed.), *Nuevas formas de docencia en el área económico-empresarial* (pp. 151–167). Edición digital @ Tres.

80. Sánchez-Vera, M. M., & Prendes-Espinosa, M. P. (2015). Beyond objective testing and peer assessment: alternatives ways of assessment in MOOCs. *RUSC: Universities and Knowledge Society Journal, 12*(1), 119–130. https://doi.org/10.7238/rusc.v12i1.2262.
81. Santoveña, S. M., & Bernal, C. (2019). Explorando la influencia del docente: Participación social en Twitter y percepción académica. *Comunicar, 58,* 75–84. https://doi.org/10.3916/c58-2019-07.
82. Seibu, M. J., Biju, I., & Yakub, S. (2006). Impact on students learning from traditional continuous assessment and an e-assessment proposal. In *The Tenth Pacific Asia Conference on Information Systems* (pp. 1482–1496). University Putra Malaysia. Retrieved from https://bit.ly/3d5 5t2L.
83. Sewell, J. P., Frith, K. H., & Colvin, M. M. (2010). Online assessment strategies: a primer. *Merlot, 6*(1), 297–305. Retrieved from https://bit.ly/2uGVKP4.
84. Sharma, D., & Karforma, S. (2012). Risks and remedies in e-learning system. *International Journal of Network Security and Its Applications, 4*(1), 51–59. https://doi.org/10.5121/ijnsa.2012.4105.
85. Shih, R. C. (2011). Can Web 2.0 technology assist college students in learning English writing? Integrating Facebook and peer assessment with blended learning. *Australasian Journal of Educational Technology, 27*(5), 829–845. https://doi.org/10.14742/ajet.934.
86. Shraim, K. (2019). Online examination practices in higher education institutions: Learners´perspectives. *Turkish Online Journal of Distance Education, 20*(4), 185–196. https://doi.org/10.17718/tojde.640588.
87. Skirpan, M., & Yeh, T. (2015). Beyond the flipped classroom: Learning by doing through challenges and hack-a-thons. In A. Decker & K. Eiselt (Eds.), *SIGCSE'15: Proceedings of the 46th ACM Technical Symposium on Computer Science Education* (pp. 212–217). Association for computing Machinery. https://doi.org/10.1145/2676723.2677224.
88. Slavin, E. (2014). Cooperative learning and academic achievement: Why does groupwork work? *Anales de Psicología, 30*(3), 785–791. Retrieved from https://bit.ly/38nBdxx.
89. Solomon, G., Schrum, L. (2010). *Web 2.0: How-to for educators.* International Society for Technology in Education: EEUU.
90. Subramanian, N. S., Narayanan, S., Soumya, M. D., Jayakumar, N., & Bijlani, K. (2018). Using aadhaar for continuos test-taker presence verification in online exams. *Advances in Intelligent Systems and Computing, 701,* 11–19. https://doi.org/10.1007/978-981-10-7563-6_2.
91. Takala, M., & Wickman, K. (2019). Collaborative case-based virtual learning in higher education: Consultation case in special education. *Journal on Digital Learning in Teacher Education, 35*(4), 236–248.
92. Trentin, G. (2009). Using a wiki to evaluate individual contribution to a collaborative learning project. *Journal of Computer Assisted Learning, 25,* 43–55. https://doi.org/10.1111/j.1365-2729.2008.00276.x.
93. Vargas-Vargas, M., Mondéjar-Jiménez, J., Meseguer-Santamaría, L., Alfaro-Navarro, J. L., & Fernández-Avilés, G. (2011). Cooperative learning in virtual environments: The Jigsaw method in statistical courses. *Journal of International Education Research, 5*(7), 1–8. Retrieved from https://bit.ly/2X2ytCE.
94. Vera, M. J. (2014). La evaluación formativa por pares en línea como apoyo para la enseñanza de la expresión persuasiva. *RED, 43.* Retrieved from https://bit.ly/38avaMM.
95. Wang, Y., & Zong, Z. (2019). Why students have conflicts in peer assessment? An empirical study of and online peer assessment community. *Sustainability, 11*(23), 6807. https://doi.org/10.3390/su11236807.
96. Yuste, R. (2013). *Una e-evaluación innovadora como factor de mejora de la enseñanza online* (Doctoral thesis, Universidad de Extremadura). Redined. Retrieved from https://bit.ly/2Zi kXfX.
97. Yuste, R., Alonso, L., & Blázquez, F. (2012). Synchronous virtual environments for e-assessment in higher education. *Comunicar, 39*(20), 159–167. Retrieved from https://bit.ly/2Tn6lbm.

Chapter 4
Online and Collaborative Tools During Academic and Erasmus Studies

D. M. Oliveira and A. L. Terra

Abstract Part of the students' academic path is the elaboration, construction and presentation of works where there is an interaction between various elements of a group. It is common, in many countries, for students to actively participate in groups, either between classmates or even from another country, with the aim of designing, creating and presenting tasks, where information should be viewed and changed by all group members, if possible, simultaneously. A few years ago, in order to be able to carry out this kind of academic group work, it was necessary to synchronize times, days and places with the group members so that the work meeting could take place. Now, this interaction is virtually possible, first with chats, videoconferences and a variety of virtual tools and their many possibilities. This chapter starts from a literature review on studies with online collaborative work platforms to analyze, through testimonials from higher education students, if online collaborative work tools have been an asset for students during their academic career, including participation in the Erasmus Program in different countries. It was found that university students find these tools useful, despite the fact that the uses do not have the support or guidance of their educational institutions and are not accompanied by an in-depth study of these tools, which have caused problematic situations that could have been avoided.

Keywords Online collaborative work tools · Life history · Google Drive · OneDrive · Erasmus Program

D. M. Oliveira (✉) · A. L. Terra
Polytechnic Institute of Porto, CEOS.PP—Centre for Organisational and Social Studies of P.Porto, São Mamede de Infesta, Portugal
e-mail: dalbertoliveira@gmail.com

A. L. Terra
e-mail: anaterra@iscap.ipp.pt

R. Babo et al. (eds.), *Workgroups eAssessment: Planning, Implementing and Analysing Frameworks*, Intelligent Systems Reference Library 199,
https://doi.org/10.1007/978-981-15-9908-8_4

4.1 Introduction

The elaboration, construction and presentation of academic works, where there is an interaction between various elements in a group, are part of the reality of most, if not all, students, regardless of school level. This practice has been developed and studied by some authors such as Gaillet [1] and Donelan et al. [2]. Without going into the merit of the discussions about whether this type of work is collaborative or cooperative, issues discussed by Torres et al. [3] and Paulus [4], group work presents advantages such as interaction and exchange of experience, which are not obtained through individual works [5, 6]. Often students actively participate in groups, whether between classroom colleagues [7], or even among colleagues in another country [8], in order to design, build and present papers, where information must be seen and changed by all members of the group, if possible, simultaneously.

A few years ago, in order to be able to carry out group work, it was necessary to synchronize times, days and places with the group members so that the work meeting could take place. This made the process more difficult, especially when students live far from the university [9]. Currently, this presence has become possible in a virtual, synchronous or asynchronous way [10], first with chats, then with video conferences and a panoply of virtual tools and their numerous possibilities [11, 12].

This chapter aims to analyze whether collaborative online working tools are indeed useful for higher education students in different countries. This objective will be achieved using the Life History Methodology, seeking a different look at the phenomenon of the use of online collaborative work platforms, from the students' point of view. Considering what has already been commented on by authors such as Kouritzin [13], Closs and Antonello [14], Nogueira et al. [15] and Wright [16]. Bearing in mind problems such as subjectivity, reliability and verifiability, but integrating the analysis of the literature with the testimonies reported, in order to confront these deficiencies.

To achieve this goal, applying Life History methodology, testimonials of students reached through the Erasmus Program will be used, which comprises an international project that seeks to provide opportunities for people and organizations, allowing the development and sharing of knowledge and experiences in institutions and organizations of different countries [17].

In the first part of the work, within the theme "State of the Art", a brief overview of academic work before and after the advent of computers is presented, the bibliography on academic work, focusing on the features of free online collaborative work tools, and these tools, namely, Google Drive [18 Google, 2020] and OneDrive [19 Microsoft, 2020] are then presented, comparatively. Soon after, Life History begin with their testimonies that address the use of these tools during the academic journey of students in different countries, through the Life History of these students, both in their academic path in their country and as part of the Erasmus Program, comparing these testimonies whenever possible with what is commented on by the scientific literature in order to understand the pros and cons of the way how this use was made by students.

4.2 State of Art

Used by teachers and students, sometimes due to mutual influences, other times due to the curricular requirements, online platforms for collaborative work increasingly attract the attention of the academic community for facilitating the sharing of information in group work. From the moment that collaborative work platforms became more and more known, the practice of collaborative writing through these platforms became more usual. But, building academic work in groups was not always that easy.

As commented by Santos et al. [9], building academic works in groups, before the advent of current technologies, such as computers and Internet access, was not easy, and one of the reasons was the attempt to organize the space–time of the members of each group, especially when the students lived far from the meeting spaces. Other problems are the agreement of parts of the works developed by different members of the group. With the advent of computers and posteriori, the Internet, the "prison" in space–time is no longer a problem, and group work could now be done remotely, with easier sharing of data and information.

Collaborative work supported by computers was the subject of Jonatham Grudin's study, still in the first half of the 1990s, in the article entitled "Computer-Supported Cooperative Work: History and Focus" [20]. Among the various possibilities that a computer already allowed was the maintenance of several parts of the work in a single location, accessible for all members of the group to consult; the possibilities of search in databases, such as digital libraries inserted in floppy disks (diskettes) and posteriori in compact disc (CD), very useful in the development of academic work, as can be seen in the works of Krubu and Osawaru [21] and Lohar and Kumbar [22]; access to text editors that made manual transcriptions of academic papers unnecessary, as a way to standardize handwriting, unless the job guidelines or the teacher prohibited the delivery of typed work on computers, which was not uncommon in the early days the use of computers for this purpose and still occurs today [23–25]; in addition, access to a range, albeit small, of digital tools for the implementation of algorithms and calculations. At the beginning of the twenty-first century, unfortunately it was still possible to find teachers reticent to the use of technologies as can be read in the article by Godwin-Jones [26] on disbelief, by educators, regarding the use of new technologies for language teaching, for example.

The Internet connection, on the other hand, made it possible to share information through communication channels such as e-mail and chats, which together with the use of telephone calls and short message service (SMS), facilitated communication during academic work group production [26]. Still in an archaic way, it was possible to find information in some online scientific databases. Some students already had computers at home, connected to the Internet and with access to these databases.

Sharing information by e-mail is still a common practice for the construction of academic work in groups, even with the advent of online collaborative work platforms. In these practices, students send work among the elements of the group so that they can continue them, and sometimes there is a convention regarding the use of different names for different versions of the works, but often what is observed is only the sending of versions distinct with the same name, which often causes confusion between versions, loss of information and consequently time.

It was and still is usual for classes to have a so-called class e-mail where materials and notes are addressed so that everyone can gain access to information [27]. In the same way, working groups created a "central" e-mail with the same purpose, that is, the sharing of the information searched among the members of the group. In this way, searches were submitted to the group's e-mail, in messages, sometimes with defined subjects, if there were previous conventions stipulated by the participants and the work was constructed, each in the space–time that most suited them, without neglect the rules stipulated by the work itself, which included formatting and delivery date.

The first launches of collaborative work platforms date back to 2005 [29–31], basically as extensions in web browsers, and these have evolved to the point that they currently have an increasing variety of features and therefore new possibilities, among which it is possible to mention the online storage of documents and folders in different formats; creation, insertion and editing of metadata in these documents and folders, to facilitate indexing and searching; creation of links to share these documents and folders, with different levels of permissions, such as viewing, distribution and editing; editing of different stakeholders synchronously or asynchronously; integrated chats and videoconferences; among other features [10, 32].

All these features have enhanced, and continue to enhance, the absence of interference related to the distances of the elements of a workgroup during its production, and these facilities have made countless authors investigate how this impacts the academic environment.

Rienzo and Han [29] comment that these tools, which have little or no cost, are being used by educational institutions to better prepare their students for a world where collaborative work is done over the Internet. The authors draw a parallel between the collaborative work provided by Microsoft platforms, Office Live (currently replaced in part by OneDrive) and Google tools (Docs, Groups and Sites). The authors also observe a positive impact on the administration of courses through these platforms and foresee additional benefits as the functionalities of these platforms increase.

Srba [33] studies the use of Google's collaborative tools, including Google Groups, Docs and Calendar, as well as the use of pages in wiki format, in a pedagogical environment involving the methodology of problem-oriented projects, with students from Aalborg University, Denmark. In this study, the author noted that the students demonstrated a certain ease in the use of the platforms, but this practice required greater preparation by the advisor. He also noted that the use of Google Calendar has saved time in organizing meetings.

Miseviciene et al. [34] contribute to a comparative view between the various Microsoft tools provided by MS Live @ Edu (currently Office 365) and Google Apps, during the collaborative work of students at Kaunas University of Technology, at Lithuania, mainly regarding the use of MS Live @ Edu in the e-learning and blended learning environment. The authors identified as essential resources the communication facilities, the collaborative editing of documents, the storage and sharing of these documents and the accessibility independent of time or space. Finally, the authors note that these technologies help both students and teachers in the educational process, but also verify that not all educational institutions are qualified to use these tools, either due to the lack of training of the participants or due to the lack of adequate equipment.

Brodahl et al. [35] focus on the quantitative results of research questionnaires in which they investigate the work done on two online collaborative working tools, Google Docs and EtherPad. This research seeks to understand whether the perceptions of use depend on factors such as sex, age, digital competence, interest in digital tools, educational settings and choice of writing tool and examines whether the tools are easy to use and effective in group work.

Through an experimental study, Sánchez and de la Muela [36] start from the elaboration of a collaborative activity with undergraduate students in Early Childhood Education, through the Google Drive tool, and observe, through the answers of an applied questionnaire to the students, who although many of them have never had contact with these tools of collaborative work, and they consider the use of these tools easy and practical for the formation of groups, without the need to be close to each other. However, at the end of the experiment, these authors comment on the need to try to understand if there are differences between the skills acquired by students who participate in these activities with online interaction and those of students who participate in the same activities in person.

Boellstorff et al. [37] comment on the online collaborative writing of the manual "Ethnography and Virtual Worlds: A Handbook of Method" [38] using tools such as e-mail, Skype and Google Docs. The authors comment that during the work, each of them felt the need to use other tools and applications to assist in the progress of the collaborative work, as these needs arose. However, they also comment that the collaborative authorship practices that they already had been essential for the writing of the manual.

Naik et al. [39] compare numerous cloud-based tools and applications in order to meet educational needs in higher education in India, related to making information available quickly and accurately, regardless of the space–time in interested parties to meet. In this study, aspects related to the sharing, availability, security and reliability of the required information are considered. These authors conclude that these tools and applications offer numerous benefits for the processing and efficient use of information.

Baiges and Surroca [40] investigated, through questionnaires, the opinions of the first-year students of the Facultad de Ciencias de la Educación de la Universidad de Lleida, on the use of two collaborative online working tools, Wikispace and Google Drive. These opinions were collected after the development of two group works, one on each platform. With this investigation, the authors point out that, among both tools worked on, Google Drive benefited from a more favorable opinion on the part of students regarding ease of use. And they also comment that it is necessary a study that investigates a relation between the characteristics of the works with the online collaborative tools in order to try to understand if there is a facility or difficulty in the development of the work with the tool used.

Olson et al. [41] develop research based on a historical review of the study of collaborative writing, covering some platforms such as SASSE, ShrEdit and EtherPad. Subsequently, the authors investigated, through an empirical study with university students, how to use the collaborative online work platform, and Google Drive, together with an extension produced for this platform, DocuViz [42], which allows a more sensitive visualization, by color, of the edits made in the document, shapes the writing between pairs, during the production of a work. The authors divide the observations into three types, those of students (co-authors of the document), managers of writing teams and teachers. It is observed that Google Docs has good attributes that favor collaborative writing and the use of DocuViz facilitates the perception of how this writing is constructed. However, he comments that additional studies are necessary in order to verify, for example, whether the size of the final document contributes or not to its quality.

Abrams [43] draws on the studies on participatory patterns by Storch [44] and Abrams [45] to conduct research with groups of first-year German language students at a US university, where he proposed to investigate the action of collaborative writing of these students in a second language, using the Google Docs platform. In this study, it was found that collaborative work groups produced texts with more propositional content and better coherence than less collaborative groups, verifying mutual assistance between students encouraging the creation of content.

Liu and others [46] investigated the use of the collaborative Google Maps tool, called My Maps, in Hong Kong to register services and activities provided locally. In this study, the authors found that the interviewed local residents consider the interactive map to be a useful resource for improving access to these services. Another successful use for My Maps was investigated by Mohan and others [47] in a study in India. In this study, My Maps was used in harmony with other tools to overcome difficulties in visiting individuals who were part of an influenza control group.

The aforementioned authors comment on some platforms that enable and facilitate collaborative work online, such as tools in wiki format [48], blogs [49], Skype [50, 51], EtherPad [41, 52], Dropbox [53, 54], just to name a few and several tools from Microsoft and Google. Among these, this chapter will highlight the two that are part of the experience mentioned here, Google Drive [18] and OneDrive [19]. Both platforms have free and paid versions, in addition to students and teachers versions' that can be accessed through educational institutions that partner with Google [55] or Microsoft [56].

Another point in common between both platforms is the possibility of working collaboratively in a synchronous way, which allows multiple users to interact in the same document at the same time; and asynchronous, when, for example, in your time, regardless of whether other users are interacting at that moment in the document, the changes are recorded for other users to be able to verify in a future moment [10]. They also allow four types of collaborative work, commented on by Robillard and Robillard [57], that is, Mandatory, Called, Ad hoc and Individual.

It is worth mentioning that these platforms did not present a novelty regarding the storage and sharing of documents online, services called "free file hosting" such as Dropbox [54], the SteekR service [58], which gave instead of F-Secure Online

Backup [59–61], Uploadingit [62], among others, already did this before the services mentioned in this chapter. and however, they did not reach as much popularity as the tools discussed here.

4.3 Google and Microsoft Platforms

In a structured interview with a university student (see Sect. 4.4) about the use of collaborative work platforms, during his academic career, it was observed that two platforms were mentioned and used by this student and his classmates. These platforms are Google Drive and Microsoft OneDrive. When this interview was extended to other university students, it was also observed citation of both platforms. In Sect. 4.4 of this chapter, the testimonies of these and other students on the use of these platforms will be addressed. For now, this chapter will comment on each of these two platforms, taking into account the features addressed by the students. Details on these features and a comparison between them can be seen in Table 4.1.

As can be seen in Table 4.1, individual or collaborative work with Google Drive or OneDrive is very sensitive. Documents can be created in numerous text editors, including free ones, spreadsheets or slideshows, to name just three options, and transferred to the platforms or extensions installed on the personal computer, either by dragging/dropping, copy/paste, or by uploading. Documents can also be created within the platforms themselves or their extensions. For others to be able to view and/or edit documents at the same time, an invitation link, with the mentioned permissions, must be sent.

There are many resources within the platforms, with the creation of groups, the use of various elements in editable documents, such as graphics, images and hyperlinks. There are also control features such as timestamp and revision history, storage, organization, data management, such as the possibility of keeping drafts in one place, checking what has been rewritten and/or erased, in addition to the possibility of being able to follow and/or check the work of each member of the group, using the editing color palette that uses a different coloring for the source of each participant [45].

It should be noted that numerous formats may be lost or even inconsistent, both during the conversion process and in the process of editing documents created in formats other than the platform [98], or even when they are used differently browser [99]. However, this loss of formatting also occurs when documents from the same platform are opened in different versions [100], so it is not uncommon to find text documents outside the expected format, because despite using the same platform of creation for the opening of the document, the version used for both processes are different.

Another issue that must be considered is the intervals between writing and storing this writing by the platform itself and making updates available to other writing stakeholders. This time lapse can be caused by deficiencies in the equipment used, the ability to transmit and receive data from the Internet and its transmission equipment such as cables or routers and even by the company providing the signal. These issues

Table 4.1 Comparison between Google Drive and OneDrive

	Google Drive	Microsoft OneDrive
Since	2012	2014
Overview	Services for creating, storing, sharing and synchronizing folders and documents in various formats, such as text documents, spreadsheets, questionnaires, among others	
History	Inserted in Google Drive is Google Docs [63], which contains a suite of tools used for collaborative writing and this suite dates to 2006, when Google acquired Upstartle [64], the developer of Writely, a processor of online collaborative text created in mid-2005 [30]	This platform started to be developed in 2005 with the purchase by Microsoft of FolderShare [65], a service that started in 2002 and that allowed the storage, synchronization and remote access of files through an Internet browser In 2007, Microsoft launched SkyDrive [66] and it has undergone several changes over time. Examples of these changes are [67–70] Finally, in January 2014, the platform is renamed OneDrive [71, 72]
Access mode	Both platforms can be accessed directly through a browser [18, 19], or installed on compatible equipment (e.g., computers, tablets, smartphones) by downloading an application [73–77]. It is also possible to access these platforms through an extension installed in certain browsers (e.g., Chrome) [78, 79]	
Types of documents that can be created and edited	Text documents, spreadsheets, slide show, surveys, notebook	
	Drawings, maps, Web sites, and with the application connection, numerous other document formats	
Example of other Apps	DocSecrets [80, 81] is a plugin that allows you to hide part of a document, from a word to one or more pages. In this way, parts considered "confidential" of a certain document can be hidden while the rest of the document is shared	The IFTTT (If This Then That) [82] platform, which allows automating tasks not only on OneDrive, but also in other applications like Gmail for example. Among these automated tasks it is possible to save a photo in OneDrive in which a certain person was identified on a social network or connect virtual assistants to perform tasks in OneDrive

(continued)

Table 4.1 (continued)

	Google Drive	Microsoft OneDrive
Free space	Documents created through the platform, that is, which are part of the standard Google Drive formats, can be stored on the platform itself without occupying the space available for storage, this space is 15 GB for recent users, using a Google account, which are divided by all Google products associated with the user account e-mail. These products include e-mail space (Gmail) [83], photo storage (Google Photos) [84], Google Sites [85], Google Drive itself and other products that the user may have on this account. Documents created on other platforms, such as a Microsoft Word document, for example, can also be stored in Google Drive, but these occupy part of the 15 GB of available space [86]	All files inserted in OneDrive take up storage space. The free storage is 5 GB [87]
Document conversion	Numerous documents created in other formats can be converted to the Google Drive format and/or edited in the original format [88]. Examples of documents that can be converted to Google format are Microsoft Word, Excel and PowerPoint [89], Apache OpenOffice documents [90, 91] or [92]	Microsoft Office Open XML Format (.docx,.pptx,.xlsx) and OpenDocument Format (.odt,.odp,.ods) [93] extensions
Document sharing	It is possible to define a public link for the documents and in this way anyone with access to this link will be able to view the document It is possible to allow only certain users (e-mails) to access the documents Another sharing option is to publish the document on the Internet, on a Web site, for example, and this can be done through the "Embed" function after publication. Through this function, the document will be available to the public and can be searched by the search engines. This function does not remove the sharing permissions previously commented [94–96]	

(continued)

Table 4.1 (continued)

	Google Drive	Microsoft OneDrive
	There are three permission options: display only, which, as the name implies, allows the recipients of these links to view only the documents; viewing and comments, which allow you to comment on the document; and, finally, an edition that allows working on collaborative writing Finally, through the "Sharing Settings" dialog box, you can allow or deny the possibility for editors to modify access options or add other people to the document, in addition to preventing people from just viewing or commenting on the document. Transfer, print or copy [91, 95]	It is possible to define permission levels for shared documents: viewing only or viewing and editing For customers who have a paid subscription (Premium), it is possible to set an expiration date for the invitation link and enter a password to access the document [87 Microsoft 2020] [97 Microsoft 2020]

of difficulties or even errors in updates have already been investigated by authors such as Ahmed-Nacer et al. [101].

4.4 Life History: Group Work After Online Collaborative Work Platforms

In order to try to understand how university students used collaborative work tools during their academic career, the Life History Methodology was used, which allowed to collect more detailed personal views, with the inclusion of a perspective of personal meta-analysis, that is students remember and self-reflect on their past practice and provide unique data that other data collection techniques, such as surveys or questionnaires, would not provide.

The Life History Methodology was used, integrating the analysis of the literature with the testimonies reported, in order to avoid pitfalls related to reliability and verifiability, as discussed by authors such as Kouritzin [13] and Closs and Antonello [14], that bring in their studies references to the works of Denzin [102] and Hatch and Wisniewski [103], while considering the study by Nogueira et al. [15] that demonstrates the attention that must be paid to subjectivity and the observation from one point of view to the detriment of others, when using this type of methodological approach; however, it should be noted that this is a methodology that allows the researcher a different look at the phenomenon, which posteriori may be confronted with either existing knowledge and/or knowledge that will still be manifest [16].

Initially, a structured interview was conducted with a student (Student A), who participated in the Erasmus Program between the years 2015 and 2020, and was interviewed, in order to understand its relationship with online collaborative work platforms, in his academic path, which included attendance at activities, both at the institution in his country and at the other institutions where he completed the international mobility period (Erasmus Program). In this interview, the Student A reported, in detail, the practices he developed and participated using these platforms. In this way, it was possible to obtain a point of view on the use of these platforms in higher education, including in academic mobility. This first interview was conducted in January 2020. From the analysis of Student A's life history data, an interview guide was designed for other students.

To increase the university students' perception using online collaborative work platforms, another 13 students were contacted in May 2020, at random, via social network, based on the contacts that Student A had with students who participated in the Erasmus Program in the same countries/time as he, without, however, having been colleagues of course/class. The difference between the date of the collection of Student A's testimonial for the other students happened because it was necessary to analyze Student A's testimonial to then build the questions that were presented to the other students (see Appendix). Of the 13 requests sent, 9 responses were obtained, of which 7 stated that they used collaborative work tools from Google and Microsoft during their academic career. Table 4.2 shows the list of 8 students, including Student A, their country of origin, their undergraduate course and the country (ies) where they were on the move.

After these considerations, the text will start to comment on these testimonies and to fit them, when possible, in the literature review previously mentioned.

Table 4.2 View of permissions on events (E) and their details (D) on Google Calendar

Student	Country of origin	Academic course	Erasmus country
A	Portugal	Degree in Sciences and Technologies of Documentation and Information	Spain, Russia, Poland, Brazil, Kosovo
B	Iraq	General medicine	Russia, Serbia
C	Austria	Bachelor of arts	Russia
D	Romania	Culture and politics in European and International Context	Poland
E	Moldavia	Management	Poland, Spain
F	Russian	Economic and Management	Czech Republic
G	Russian	Linguistics	Czech Republic
H	Egyptian	Architecture	Russian

Source The authors

Table 4.3 General testimonials on the use of collaborative work tools

Student	General comments regarding platforms
A	"As soon as I entered higher education, I started to use the Google Drive application to store the files made available by teachers and the Google Docs application to take note of what was said by them. I organized the content by creating folders inside Google Drive and kept an offline copy of all the content"
B	"These tools had positive effects on me and the rest of the students because it helped in a lot of things and made it easier for the student, such as storing information and saving sources and documents [...] I think Learning is no longer limited to classrooms. The advent of new technological tools has made it possible for distantly located students to collaborate with their instructors and peers for learning new skills and acquiring enhanced knowledge. Distance and time are no longer a barrier for imbibing knowledge"
C	"We mainly used different tools of Google Drive in Austria, as it was preferred by students as well as by teachers"
D	"An important aspect of using Google Drive is that you can have a backup of your work and for a long-term group project I think it's essential and you avoid unpleasant situation of losing your files and start over"
E	"I think these tools definitely give you the ability to work efficiently while other can participate at it without any restrictions, limitations, and other issues" and keep saying "No time limitation, easy- access, the opportunity to share and edit, safe place for uploading sensitive content and so on"
F	"It's saved time for everyone and has common tools also [...] Several people can edit in the same time"
G	"We mostly used it for creating presentations/calculations so as several people could work simultaneously"
H	"Actually, back in our university we used to share only files by Google drive and the other tools we did not used at all as our major depends more on individual projects"

In Table 4.3, it is possible to read a summary of what each student comments, in general, on the use of collaborative work platforms:

In these testimonies, it is clear the usefulness of online collaborative work platforms, for the storage and management/organization of study material, for the editing of these contents, for the monitoring of classes and for the consultation and subsequent reissue of this material, even without access to an Internet connection.

However, for these platforms to fulfill the role of an online collaborative work platform, they must facilitate group work between different elements, whether these are Mandatory, Called, Ad hoc or Individual works [57], ensuring that all these elements have access to the set of tools and features such as the storage, retrieval and editing of information, whether synchronously or asynchronously [10], by several elements in a working group. The next subtopic will address examples of how these works were constructed.

4.4.1 Completing Assignments Through Free Versions of Online Collaborative Work Platforms

The questions below were taken and translated from statements of work made available by teachers in the Moodle of the Degree in Sciences and Technologies of Documentation and Information, course attended by Student A:

> In a group of 3 to 5 colleagues, research about;
>
> Answer the questions below … in groups of 5 members;
>
> Get together in groups and investigate …;
>
> Based on studies A, B and C, get together in groups of X members and elaborate….

There are many times when guidelines such as those described above are inserted in academic works.

These sentences relate to the gathering and sharing of information, in a group, with the intention of reaching a certain objective, and these practices have challenges related to the way information will be acquired and selected, a subject addressed by some authors such as Talja [104]. However, these challenges bring with them a set of benefits such as collaborative writing already covered earlier in this chapter (see, e.g., [41]).

With these sentences in mind, the next report addresses how the use of an online collaborative work platform was helpful in solving the proposed activities during the course:

> One of the first academic works that we carried out was the resolution of a worksheet with seven questions that should be answered by a group of five elements. As a source of consultation, we had the slides present in the discipline's Moodle, a book in virtual format made available online and it was also allowed to consult other sources that the group members considered relevant.
>
> After the formalization of the group, which was built by affinity between ourselves, as students, I commented with my other colleagues about the existence of Google Drive, which I used and would like to test it with everyone, to resolve the proposed work.
>
> The other colleagues agreed, so I transcribed the form to a Google Docs text editor sheet using the copy/past resource, collected the email from the other members of the group and sent them, from Google Drive, a link from text editor sheet with editing permissions. (Student A)

In this testimonial, Student A comments on the existence of a worksheet that should be answered in a group and that to solve this work, it was necessary to search for information, one of the sources for carrying out this research being the Moodle platform. This platform, in addition to serving as an information repository to support teaching and a tool for modeling courses, is also a platform that enables collaborative work online synchronously and asynchronously [10, 105].

However, it is possible to observe in this provided testimonial that Student A prefers the use of another platform, extra Moodle, Google Docs. It is not clear from this testimonial why Student A prefers Google Docs and not another platform like Microsoft's OneDrive. To use it, Student A first advocates about this platform with his colleagues in the group and having their approval the student proceeds to insert the questions of this work on the Google Docs platform and use the functionality present on the sharing platform with editing permission so that all members of the group can have access to the document.

The resolution of group work and other facilities such as the copy/paste functions of any platform for Google and Microsoft platforms and the compatibility with numerous types of documents were points mentioned by other students, as can be read in Table 4.4.

During the collection of these testimonies, in first semester of 2020, a pandemic struck a large part of humanity, causing disruptions in various sectors of society, including the academic [106]. School progress, group work and even literary productions despite having been greatly affected were helped by the use, by teachers, students and researchers, of collaborative work platforms such as Google Drive and OneDrive, as can be read in the testimonies of students to follow:

> Especially because it's quite hard to get together during this situation [the pandemic], Google drive is a perfect platform for teachers who upload materials for the students because of the new way of learning (online system)." […] For example now I have all the courses online [in Spain], due to all the teachers have to adapt to a new way of teaching. In this case have experienced moments when the teacher would upload several screen recordings of the lecture, tasks, or material on Google text editor, or spreadsheet, in that way that all the students can easily access the materials. (Student E)

> During the pandemic I had to work on articles together with colleagues and teachers and we used OneDrive for this purpose. (Student A).

However, as these tools are still unknown to some students, difficulties and errors occurred, such as the lack of perception about the need to have an e-mail compatible with the platform, tests with different browsers (Cf [99]) or lack of understanding in the collaborative use of spreadsheets. Table 4.5 illustrates some of these "problems":

Table 4.4 Testimonials about the ease of using collaborative work tools

Student	Comments on facilities on the Google platform
B	"These tools helped me in my studies a lot to copy, store and preserve the information thereof save and share documents in Google Drive, Google Drive text editor and […] OneDrive"
D	"One of my favorite facility is the possibility to open and edit various kinds of documents and I don't need a special software anymore; I just upload it and open from Google Drive"
E	"I think Google is such a powerful platform that these tools just make it even better. Sharing picture, deposit pictures, sharing projects, editing the restrictions and limitations for those who you don't want to see your page on Google drive definitely have a good impact […] Google drive offers me the opportunity to work with other colleagues on the same project while sharing every step online […] Plus I found it quite useful for students or people who always have projects in a team, group format and have no time or possibility to get around for getting it done"

Table 4.5 Testimonials about the difficulties in using collaborative work tools

Student	Reports on difficulties in using tools
A	"Some students have an initial difficulty in editing the documents because they did not have an email registered with Google and the initial configuration of the link did not allow anyone with the link to edit the document, if not only the owners of the registered email addresses on Google. After this initial difficulty and the links were resent, all elements were able to access the Google text editor and started viewing the worksheet issues and working on their resolution"
D	"When I was working on a group project using Google Drive spreadsheets some of my colleagues were not used to this tool, which made collaboration difficult […] if I use Mozilla Firefox makes it more difficult to upload files"
F	"You need to have Google account, so need to register"

After solving the initial problem with emails, Student A continues his testimonial exposing the platform features used to solve a specific job, in addition to commenting on other difficulties inherent in this resolution:

> Among the resources used in works I can mention the chat integrated to the platform, which can be accessed in the document that was being edited. In this chat we exchanged information about academic work and asked questions about the platform used, one helping the other. Through the chat we agreed to insert the hyperlink of the sources consulted as footnotes to the answers that were being prepared.
>
> The difficulty in this work, in particular, was related to the fact that the answers to the questions were interconnected and thus the construction of the answer to each of the questions depended, in part, on the answers of the other questions. Through the platform, everyone could see the construction of the answers live and so each of us could work on our own resolutions. Everyone also had direct access to the sources consulted so that they could confront possible differences.
>
> In this work, success was achieved, the final document was formatted according to the required standards and transferred in Word format to be delivered on the Moodle platform, used by the educational institution. However, we found that after the transfer the formatting had been modified, but the information remained unchanged. (Student A)

In this excerpt from the testimonial, positive and negative aspects of this collaborative work platform are evident. Among the positive aspects, it is worth mentioning the possibility of dialogue between the elements of the group within the platform, without this dialogue mixing with the answers to the work questions, that is, the platform has an integrated chat that allows dialogue with the stakeholders. It was commented that the footnote feature was used to share links from the sources surveyed, despite the existence of the side comment feature. As a negative point, some formatting was lost when the document was transferred to Microsoft Word format (see [36, 40, 41, 98, 99]).

Student A, still referring to the same academic work, continues his testimonial.

> In this work, in particular we were inside the same classroom, next to each other, but this experience served to realize that the same work could be done regardless of the physical location where we were, requiring only equipment compatible with the platform and an internet connection. (Student A)

From the moment these elements of the group discovered the benefits of platforms like Google Drive, a collective process began, called by Mike Murray as evangelism that consists in influencing others to use digital tools that aim to facilitate the execution of activities. This process begins raising awareness among individuals, encouraging them and guiding them in relation to the tools. In the previous testimonial, the tools by which students evangelized are those available on Google Drive; however, as will be seen in subsequent testimonials, Microsoft tools also began to be used as academic work became more robust [107] (cf [108]).

This evangelism was also observed in the testimonies of other students: "*I was advised by my colleagues or friends to try using Google drive, [...] as it is so much easier to see others work meanwhile you can edit it anytime, especially these tools are good for team projects or group homework*(Student E)". And Student G: "*in the Czech Republic [these tools], they were preferred by teachers*".

In a work considered "more technical", Student A makes the following comment about the application of Google Drive tools:

> In a second, more technical work, a thorough investigation of a topic and the construction of a report regarding this investigation was required. There were no objective questions to be answered, but on the contrary, it was expected to perceive with some depth the subject investigated. This report should be divided into parts and it was these that guided the division of this work by the members of the group. The final report was due at the end of the semester. The lack of experience with the platform and letting us guide only by intuition, produced numerous errors that forced several reconstructions.
>
> A positive point of these platforms is the fact that it is possible to work on the document regardless of time or space, as well as working on the possibility that all elements of the group worked simultaneously, in a collaborative writing. However, working simultaneously is related to the fact that those who are working on the document, are online at that moment and not working offline (which is also an option on the platform). However, what happened was many mismatches and loss of information, because members of the group often downloaded the document and worked in isolation on its versions and when uploading to the platform, mismatched information occurred. At other times, there were flaws in the internet signal and the document, supposedly online, was not being updated as expected and the confrontations of information that should be analyzed, were ignored by elements of the group, causing conflicts and loss of information.

It is interesting to note that the lack of knowledge on the part of students about the functionalities of this collaborative work platform did not prevent its use. However, the reported problems could have been minimized by searching tutorials on the platform itself or reading articles on this platform, such as the work of Ahmed-Nacer et al. [101] that address all these problems by investigating some flaws in both the Google Drive platform and Microsoft's OneDrive (formerly SkyDrive), even pointing out some possible improvements.

In Table 4.6, it is possible to read what some students commented on the way they learned about the platforms, some researched about them on the Internet, and others learned through attempts.

After this learning, during use, having been consolidated, even partially, the work could continue without many interruptions, as Student A continues:

Table 4.6 Testimonials about learning to use collaborative work tools

Student	Comments on how they learn about platforms
B	"I always search in the Internet and watch videos to gain information about the tools"
C	"I did not do research, it was mainly learning by doing"
D	"[…] first I searched some tutorials to learn how to be used it. I also researched storage capacity and data protection […] In my opinion, you must first experiment by your own how it works and then use it in group projects"
E	"I haven't [study about it]. I think using them is intuitively, you learn by exploring the platform"
H	"I prefer using YouTube or Udemy for learning any new tool"

> It took some time for everyone to realize the need to check their connections before closing the document and under no circumstances remove them from the internet and put them back on without first checking the changes in the other elements.
>
> We also noticed that it was not worth insisting on formatting the document in the platform format, because it was, most of the time, lost during the transfer of the document to the Microsoft text editor format, which most of us used and which was also installed on the college's computers. Therefore, we started to leave the final formatting of the document to the end of the work, after we have already done the transfer with the Microsoft Word extension to submit it in Moodle.

After the students understood some aspects of the work with the Google platform and started to become familiar with it, they started to use other tools of this platform to facilitate other works as can be read in the testimonial that Student A provides:

> In another academic work, the class was asked to interview a professional who graduated from our course and was working in the training area, in order to collect his professional life history so that we could relate to his training and work.
>
> For the construction of this work, we used the forms of Google Docs collaboratively. Likewise, as with sharing a text editor sheet, the same was done with a Google form. Each member of the group was responsible for a group of questions that would be inserted in sections of the form in order to enable the construction in isolation of each part of the form. In the end, these sections were linked using the form's functionalities and the final product was submitted by email to the previously chosen professional, using a professional social network called LinkedIn

In the testimonial cited above, the use of Google's forms tool is observed. This tool has a number of features, and some of these were mentioned in the testimonial, such as the possibility of collaborative construction of the form, the creation of independent sections, the ability to join these sections and finally the possibility of sharing through a hyperlink, to filling it out.

As it could be read in a previous testimonial, the sources of information searched for the resolution of working in groups were inserted as footnotes in the elaborated response texts. However, over time, students came to know other tools and features for collaborative work, as commented by Student C: "*The possibility to share and edit documents on Google was the best way to work on a glossaries or documents at the same time without making the same changes for example*".

Student A comments on another tool used:

> At this time, we already mastered, as a group, numerous resources on the Google platform, we already used the resource of suggestions for editions and supplements such as Zotero, for the creation of bibliographic references [see Fig. 4.1]. Some students started to install the Google Drive folder on their computers, so it was possible to synchronize documents from Drive directly on the computer and access them even without the need for an internet connection. However, it was very clear that if the intention was to edit these documents, together with others, it was important to always check the maintenance of the internet connection, because if simultaneous editions were being made by other people, in the same document, conflicts would occur between information, which often caused loss of information.

Review features, such as editing suggestions and the installation of add-ons like Zotero, began to be used by students. Zotero, available on the Google Docs platform since the second half of 2010 [109], is a useful tool for creating and maintaining bibliographic references [110]. Figure 4.1 illustrates how the use of editing suggestions and Zotero were used by students.

Among other tools used by the students in question is the Google Calendar and the tools for sharing and collaborative editing of events, as can be read in the following testimonial:

> We used the academic calendar to synchronize the dates of the teaching activities (classes, exams and assignments) with Google Calendar. I took the initiative to create a timetable for the class on this platform and shared the link to it with students interested in using the tool. Everyone who received the call was able to add information to the calendar and edit the

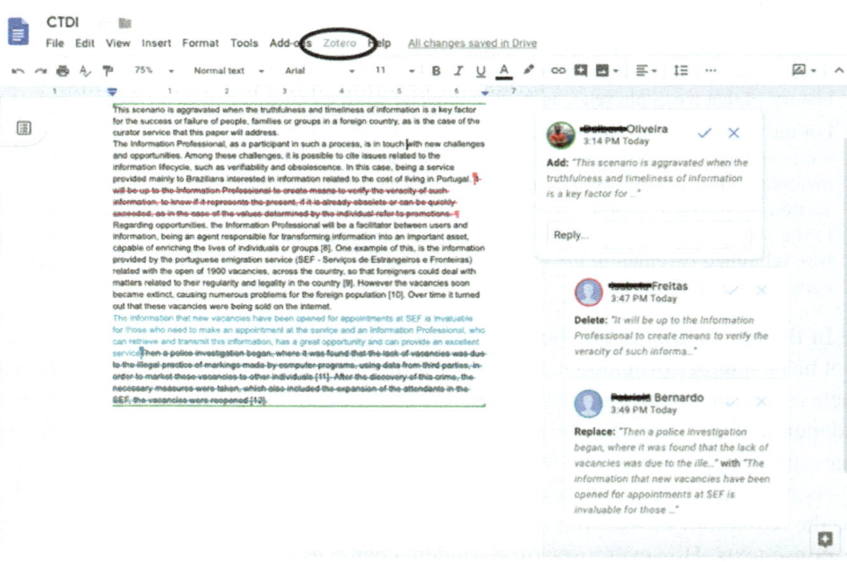

Fig. 4.1 Screenshot of the Google Docs text editor where you can view the editing suggestions on the right side of the screen and the Zotero supplement inside, circled at the top of the image. *Source* Google Drive

> information entered there. This information was related to the days, times, location (room number) and description of activities. In the description, we insert the statements of the works, or the parts that should be studied for an exam.
>
> Still on the calendar, we defined specific colors for each type of activity. In this way, each of the different classes had a different color, activities related to work could be visualized in gray and activities related to exams viewed in red. We created a set of alerts, in predefined days and times, related to the date/time of the activity, that is, we schedule the receipt of alerts with a few days and the repetition of it with a few hours before the tests and with periodic reminders about the delivery of work or parts of it. Each set up personal alerts using the calendar application on their smartphone and/or tablet.
>
> Finally, some students made use of the Google Calendar layers, which allows each user to overlay several calendars, such as the class calendar with one or more personal calendars, for example [see Fig. 4.2]. (Student A)

As it could be read in the testimonial, Google Calendar proved to be useful to keep students notified and alert about teaching activities, and it used features such as sharing the event with editing permissions, the use of colors to facilitate the visualization of different activities, the scheduling of reminders with different dates and times in order to collaborate for the preparation for exams, assignments and other activities and, finally, the possibility of overlapping several calendars, which allowed checking, for example, if events on one calendar could not overlap events on another calendar (Cf [33]).

Other students commented on the use of Google Calendar:

> I used these tools because it's an easier way to work in group, especially when we didn't have a place to meet or when we had different schedules" (Student D). And "it has cool tools

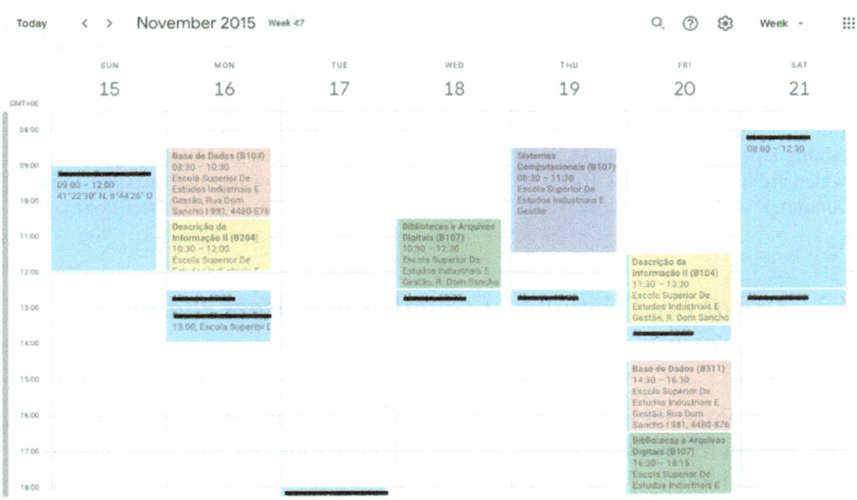

Fig. 4.2 Screenshot of the real calendar used by several students in the 3rd semester of graduation. The events in blue correspond to events in one of the personal calendars, superimposed on the class calendar. *Source* Google Calendar

> as sharing calendars, projects with others while they can edit whenever or whatever they want. (Student E)

After advancing semesters, students began to face difficulties in maintaining the use of the Google suite for collaborative work to solve academic tasks, because, according to the students themselves, the Google platform started to present problems when the academic task required more robust formatting of documents and work with bibliographic referencing:

> As semesters progressed and as academic work demanded more robust formats and references in accordance with technical standards such as the Portuguese Standard (NP 405) and the American Psychological Association (APA) standard, the Google platform began to present problems, as Microsoft's communication with Microsoft documents was very poor.
>
> Due to these problems, we gradually migrated the works to the Microsoft platform, OneDrive. As many of us used Microsoft's Office 365 package, which is offered free of charge to students at our educational institution and due to previous practice with Google's collaborative working toolset, the development of work on the OneDrive platform proved to be much more sensitive. (Student A)

These are not new problems; in the literature, it is possible to find these same problems [101]. However, the use of the Microsoft suite proved to be more sensitive regarding the maintenance of formatting and the work with the construction and updating of bibliographic references, despite presenting the same problems of synchronization between versions of the same document, when due account was not taken care regarding the maintenance of the document online.

Student A continues:

> After a new period of evangelization, which includes influencing each of the students during the preparation of works, it could be observed that some students already had the OneDrive folder installed on their personal computers. But it was also noted that OneDrive had deficiencies similar to Google Drive, that is, we had to pay attention to the internet connection when editing documents.
>
> But unlike the Google Drive folder, the OneDrive folder installed on the computer allows editing a document without the need for a browser, directly through Microsoft programs, which are included in the Office 365 package, for example, if they are also installed on the computer, such as text editor (Word), spreadsheets (Excel) or slideshows (Power Point).

4.4.2 Use of Free Versions of Collaborative Work Platforms in Erasmus

The Erasmus Program is an international project that offers opportunities for people and organizations, allowing the development and sharing of knowledge and experiences in institutions and organizations in different countries [17]. Being in Erasmus is a challenge that allows opportunities for students who can see or even create them. These opportunities are often within reach of a click on certain nodes on the web that is the World Wide Web. Perceiving these nodules is like panning for gold, as Edward Snowden comments in his autobiography "Permanent Record" [111]. In this

Table 4.7 Testimonials on the use of collaborative work tools in different countries

Student	Testimonial on the use of collaborative work platforms in different countries
C	"It did not affect my academic work and classes as the University in Russia did hardly use any electronic devices. However, I shared pictures with students I met in Russia via Google Drive [Google Photos]"
D	"I used these tools both in Romania and Poland in order to work easier in group projects […] I think these tools have a positive impact on academic work. For example, when I studied in Poland, the weather it was bad all the time and group meetings were difficult. Google Drive tools helped us a lot. […] Another great facility is that I can share photo albums with my friends"
E	"In my case I think Google photos is quite a safe space to upload my photos"
G	"The use of such tools is underestimated in Russia (at least in the University—or may be even at the faculties—where I have studied). Neither teachers nor students recommend them for use. The reason could be the individualistic approach to performing work as the majority of tasks should be done individually by every student (at Linguistic faculty I didn't have a single group work). That's why my experience of using them are related to 1 year of study in Czech Republic […] Google drive made it easier to create group projects. Even though groups were mixed all the time, the only thing we needed to start working was to share our emails"

program, the challenges can be even greater than those found in the student's home institutions, often due to cultural differences.

In this part, the chapter will demonstrate, from the testimonies of students, how online collaborative work platforms facilitated the exchange of knowledge during their academic life in international mobility, and more, it can be seen that the use of collaborative work platforms does not it is used in all higher education institutions, and according to the testimonies addressed here, this non-use is related to the institution's country.

Table 4.7 shows the testimonial of some students on the use of collaborative work platforms in different countries.

Student A, during one of his mobilities under the Erasmus Program, comments on the use of Google Calendar by students at the institution where he moved in mobility. By participating in this Erasmus Program, the student can perceive in practice a wide use of the features of this tool, such as the layered functionality for the overlapping of calendars and the privacy settings in each of the layers.

> One of the tools I had contact with during the Erasmus program was the Google calendar. Students used several overlapping calendars, using the "layer" feature, which allowed overlaying these calendars in order to allow an overview of the times of the events inserted in these calendars.
>
> Each calendar layer had different levels of permission. There was a general calendar where information about events at the level of the entire University was inserted, which included holidays, festivities, reception of new students among other events; in this calendar the details of the event were shared with everyone who had access to the calendar. In another layer, there was a calendar with events related to each of the Faculty that comprised the University so that everyone could see if in a certain timetable there would be an event in a certain Faculty, however only individuals related to that Faculty had access to the details

> of this Faculty. event. Thus, progressively, calendars were being built and shared through departments, courses and classes.

This use of layers with different privacy settings, which Google Calendar allows, can be seen in Table 4.8:

As can be seen in Table 4.2, only the details of the events related to the university as a whole were visible to all, but as these events became more specific, from one of the Faculties to the respective classes, the details of the same became available only for individuals directly related to these spaces.

The calendars facilitated the scheduling of non-conflicting meetings between the participants, facilitated the monitoring of events and the sending of reminders about these. As noted by Srba [33], the use of calendars reduces the time in organizing meetings.

In another participation in the Erasmus Program, Student A was faced with a challenge involving the promotion of a philanthropic organization in a politically rigid country. This philanthropic organization could hold its meetings in the country, but it could not do any kind of dissemination, such as campaigns to attract members. However, nothing prevented her from visiting the home of people who requested such a visit. However, how to optimize this visitation work in order to reach as many people as possible with the use of few resources?

This optimization was achieved using a collaborative feature of one of Google's free tools, Your Places on Google Maps [112]. Your Places is a tool that allows you to work with maps, individually or collaboratively. It is an interactive Geographic Information System (GIS) [46, 47] where it is possible to create maps, insert and configure data, information and objects and even create a database of locations with the same functionality as Google Maps, which includes the possibility of measuring distance between points, inserting means of transport and sharing information with others. Finally, Your Places allows the same sharing settings with editing settings as other Google collaborative work tools like Docs.

This GIS was used by Student A, to build a database that served to exchange information about members of this organization. The student starts to comment:

> I got a meeting with the local presidency of the organization and proposed the creation of a map of members, among those who would be willing to informally disclose the organization

Table 4.8 View of permissions on events (E) and their details (D) on Google Calendar

	University		Faculty		Department		Degree		Class	
	E	D	E	D	E	D	E	D	E	D
University	X	X	X		X		X		X	
Faculty	X	X	X	X	X		X		X	
Department	X	X	X	X	X	X	X		X	
Degree	X	X	X	X	X	X	X	X	X	
Class	X	X	X	X	X	X	X	X	X	X

Source The authors

> with their neighbors, relatives and friends. This map would allow a geographic visualization of the area that could be worked on, in order to channel the efforts of visits to these locations and schedule these visits in order to make the most of human, temporal and financial resources.
>
> Using Your Places, a general map was created and a link that allowed the editing of information (data insertion) was shared with the members willing to participate in the project, these members in turn were instructed to insert their addresses on the map following a certain normalization. Whenever one of these members had a person who would like to receive a visit from the organization, that member marked the home of that interested person on the map, inserted the necessary observations as the best time for the visit, and/or more sensitive issues involving the security of the parties, such as the presence of people less receptive to the organization in the region. Finally, he communicated via chat on the platform itself, or through other communication channels such as WhatsApp, with the group responsible for the visit in that region. It was possible to see details about this group in the map region itself.
>
> This map was composed of layers, the places of interest were marked by balloons, which is the object used by the platform to mark.
>
> In one of the layers, called the "Friends" convention, information was inserted regarding the members of the organization who were willing to collaborate with the dissemination of the same and by convention it was decided to use green balloons. These members in turn inserted information in the layer "Interessed" (blue balloons) with the information of their acquaintances, who were interested in receiving visits from the organization. In another layer, "Area Manager" which corresponded to the red balloons, those responsible for each of the regions inserted their own information in a facilitate contact with those responsible for visits.
>
> By clicking on each balloon, it was possible to view the information necessary for the work that should be performed. It was possible to trace routes between points of interest using the means of transport recorded on the map, such as car, on foot, or public transport and also measure distances between these points.

An example of this map commented by the Student A can be seen in Fig. 4.3—which, for the organization's safety, only illustrates the map created, not representing the geography of the organization's performance, not even the country.

According to the Student A, this work still continues to work not only in the region where it was implemented but also in other regions of the country. In this way, the organization can optimize its work of publicizing and attracting new members without inflicting the laws of the country where it was inserted.

4.5 Conclusion

In this chapter, it is noted that the advent of computers, academic works, was enhanced through searches in databases on floppy disks and posteriori on CDs. Computers yet facilitated the writing process, as they not only standardized handwriting but allowed spelling and grammar errors to be corrected more easily. The dependence on synchronizing the time and the place where each member of the group was also reduced, since all the work could be centralized in one piece of equipment and could be accessed when the elements could do so, respecting both the time of the task itself and the hourly availability of equipment, or they could also be copied to diskettes

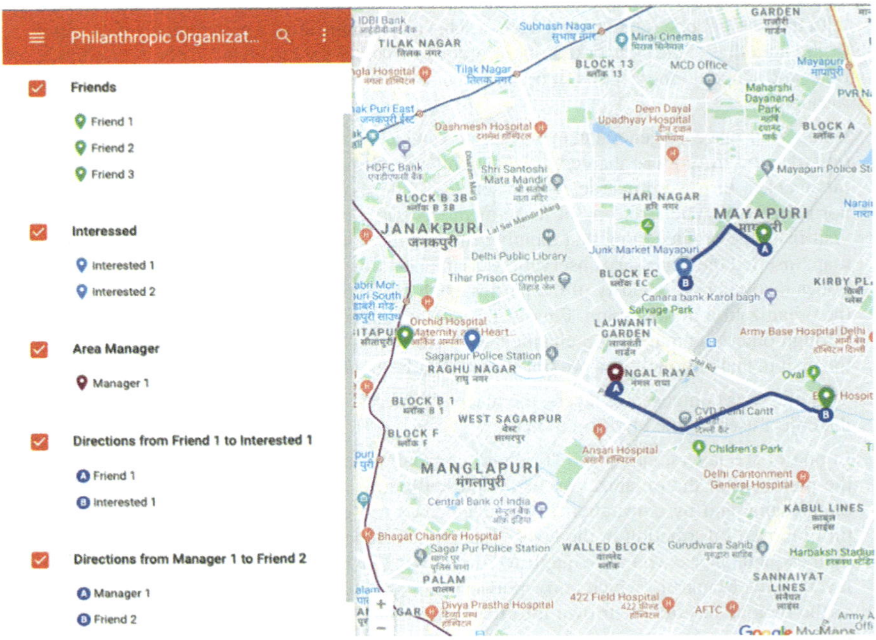

Fig. 4.3 Screenshot of a succinct example of the map created for the organization. *Source* Google Maps

and CDs for individual work on other equipment with a more convenient location for the members of the group.

From the moment that the Internet started to be used more, new possibilities focused on the accomplishment of academic activities and the possibility of accessing and sharing the activities under development through e-mails and conversations via chat further reduced the dependencies of a common physical space for the production of these activities. Finally, online collaborative work platforms like Google Drive and OneDrive began to emerge in mid-2012. Numerous authors have investigated its impact on teaching and some internal problems, related to synchronization, updates and formatting of documents during the performance of collaborative work.

As could be read during the chapter, these two platforms were used by students from different countries during their academic career and also during Erasmus mobility, including to assist a philanthropic organization in a country that is difficult to deal with politically. It is verified in this chapter that the use of these online collaborative work tools made possible the shared and simultaneous editing, online recording and auto-backup, in various types of documents such as text documents, spreadsheets, presentations, surveys, maps and calendars. These tools also made it possible to reduce temporal, human and financial resources, during the preparation of works and the practice of activities, while helping to enrich the information contained in these activities and works.

As commented by some students, these tools should be used in education as it would facilitate the academic path: "*I believe that modern technology should be included in the methods of academic study, given that such tools are free, easy to use and have various functions to make your work easier*" (Student D). And another student: "*Actually, it has a positive impact on the education process as it keeps the important files for the students and for the future ones too*" (Student H).

Returning to the objective of the chapter, which was to demonstrate whether online collaborative work tolls are really useful for higher education students in different countries, it is concluded that despite some problems that may arise, if certain guidelines of the platforms are not followed, but if the literature on the subject is taken into account, one can "generalize" that online collaborative tools, such as Google, are useful in academic works. These online collaborative tools also contribute to the reduction of temporal and material resources since it reduces the needs of these travel expenses, for example. And they become an empowering tool during times when group work cannot be done in person, as one student commented: "*How important is to find a way to work in groups even when we are not allowed to meet face-to-face*" (Student D). Even students, who did not use them in their original faculty, took advantage of them in their international mobility. However, from the testimonies, it was also possible to notice that many students learned to use these tools through trial and error, even though there are numerous tutorials on the platforms. Nor was any type of guidance, on the part of universities, observed for the training of these students on these platforms.

With this study, it was possible to expand scientific knowledge about the use of collaborative work tools online, by students, during their academic career in different countries. In this way it has relevance for several actors, it is relevant for students because it demonstrates through testimonies that online collaborative work tools, despite being used, require a certain study of their manuals; for teachers because it broadens their understanding of how students are interacting with each other, in the construction of academic work, today; and for educational coordinators because they demonstrate informational and technological needs that can be met by the school, including training on collaborative online work platforms.

However, it is worth noting, as already mentioned by some authors, what is the impact of the same activity that can be carried out using traditional methods, prior to the online collaborative work platforms, or even before the advent of computers, and by the same activity being done on a collaborative environment of online platforms? How can these tools be used on the evaluation of the working group? And yet, how online collaborative work tools can be used to assist the teacher in the assessment of individuals in a work group, while providing a fair individual assessment, according to their performance in the development of the work?

Appendix—Questions Sent to Students

1. What is your Nationality (ies)

2. What is the country of your main undergraduate education institution
3. Name of your main undergraduate course
4. Country (ies) where you studied between 2016 and 2020 as an undergraduate
5. University (s) where you studied between 2016 and 2020 as an undergraduate
6. Thinking about Google Drive and OneDrive. During graduation, which of the tools below did you use together, with colleagues, to follow classes and/or do group work? If in addition to those mentioned you used other (s) quote them.
   Save and Share documents in Google Drive.
   Google Drive text editor.
   Google Drive spreadsheets.
   Google Drive presentation editor.
   Google Drive forms.
   Google Drive calendars.
   Google Drive maps.
   Saving and sharing documents on OneDrive.
   OneDrive text editor.
   OneDrive sheets.
   OneDrive presentation editor.
   OneDrive forms.
   OneDrive maps.
   OneDrive calendars.
7. Why did you decide to use these tools? Was it encouraged by other students or by teachers? Or for some other reason?
   (You can quote individual tools if you prefer).
8. Did you search for information, such as tutorials, to use the tools? Comment.
9. In the scope of Erasmus/International Mobility, whether as a mobility student or a student who lived with mobility students, how did these tools affect (in a positive or negative way) classes and academic work? Can you comment on any examples?
10. What facilities did you find in using these tools?
    (Give examples, naming the tools, if possible).
11. What difficulties did you find in using these tools?
    (Give examples, naming the tools, if possible).
12. Found errors, in the tool itself, during its use. Comment.
13. Please write down anything you think might be of use to this study.

References

1. Gaillet, L. L. (1994). An historical perspective on collaborative learning. *Journal of Advanced Composition*, 93–110.
2. Donelan, H. M., Kear, K., & Ramage, M. (Eds.). (2010). *Online communication and collaboration: A reader*. Abingdon: Routledge.

3. Torres, P. L., Alcantara, P., & Irala, E. A. F. (2004). Grupos de consenso: Uma proposta de aprendizagem colaborativa para o processo de ensino-aprendizagem. *Revista Diálogo Educacional, 4*(13), 129–145.
4. Paulus, T. M. (2005). Collaborative and cooperative approaches to online group work: The impact of task type. *Distance Education, 26*(1), 111–125.
5. Damiani, M. F. (2008). Entendendo o trabalho colaborativo em educação e revelando seus benefícios. *Educar Em Revista, 31,* 213–230.
6. Coffin, P. (2020). Implementing collaborative writing in EFL classrooms: teachers and students' perspectives. *LEARN Journal: Language Education and Acquisition Research Network, 13*(1), 178–194.
7. Orvalho, L., Alonso, L., & Azevedo, J. (2009). Estrutura modular nos cursos profissionais das escolas secundárias públicas como trampolim para o sucesso:… dos princípios de enquadramento curricular e pedagógico… Às práticas na sala de aula e trabalho colaborativo.
8. Fernandes, M., & Santos, L. (2015). Educação para a cidadania global: Trabalho colaborativo internacional baseado em plataforma digital. *Revista De Estudios E Investigación En Psicología Y Educación, 8,* 110–114.
9. Santos, P., Fonseca, S. J., & Alvarenga, K. B. (2012) Uma análise das condições de alguns estudantes para cursar uma disciplina na área de tecnologia. In *VI Colóquio Internacional "Educação e Contemporaneidade"*. São Cristovão: Ceará.
10. Malik, M., & Fatima, G. (2017). E-learning: Students' perspectives about asynchronous and synchronous resources at higher education level. *Bulletin of Education and Research, 39*(2), 183–195.
11. Tarouco, L. M. R., Silva Moro, E. L., & Estabel, L. B. (2003). O professor e os alunos como protagonistas na educação aberta e a distância mediada por computador. *Educar Em Revista, 21,* 1–16.
12. Aguiar, A. (2012). Ensinar e aprender à distância: Utilização de ferramentas de comunicação on-line no ensino universitário. *Revista De Ciências Agrárias, 35*(2), 184–192.
13. Kouritzin, S. G. (2000). Bringing life to research: Life history research and ESL. *TESL Canada Journal*, 01–04.
14. Closs, L. Q., & Antonello, C. S. (2011). O uso da história de vida para compreender processos de aprendizagem gerencial. *RAM. Revista De Administração Mackenzie, 12*(4), 44–74.
15. Nogueira, M. L. M., Barros, V. A., Araujo, A. D. G., & Pimenta, D. A. O. (2017). O método de história de vida: A exigência de um encontro em tempos de aceleração. *Revista Pesquisas E Práticas Psicossociais, 12*(2), 466–485.
16. Wright, J. S. (2019). Re-introducing life history methodology: An equitable social justice approach to research in education. In *Research Methods for Social Justice and Equity in Education* (pp. 177–189). Cham: Palgrave Macmillan.
17. European Commission (s.d.) What is Erasmus+?. Disponible in https://ec.europa.eu/programmes/erasmus-plus/about_en.
18. Google (2020) Access all your files wherever you are. In Google Drive. Disponible in https://www.google.com/drive/.
19. Microsoft. (2020). Save your files and photos to OneDrive and get them from any device, anywhere. In *OneDrive*. Disponible in https://onedrive.live.com/about/en-us/.
20. Grudin, J. (1994). Computer-supported cooperative work: History and focus. *Computer, 27*(5), 19–26.
21. Krubu, D. E., & Osawaru, K. E. (2011). The impact of information and communication technology (ICT) in Nigerian university libraries. *Library Philosophy and Practice, 2011,* 9–18.
22. Lohar, M., & Kumbar, M. (2008). Use of CD-ROM and internet resources by the students in JNN College of Engineering Shimoga: A survey. *SRELS Journal of Information Management, 45*(2), 235–242.
23. Mesquita, M. (August, 7th 2016). Processo civil: Avaliação parcial 1. In Google Groups. Disponible in https://groups.google.com/forum/#!topic/unisaldireito2013/YSD8DSRa1EE.

24. UNITAU (2019) Trabalho 2—Processo Civil Coletivo. [Trabalho Acadêmico]. In StuDocu. Disponible in https://www.studocu.com/en/document/universidade-de-taubate/direito-civil/lecture-notes/trabalho-p-coletivo/6159333/view.
25. Pinto, F. (2012). Informática Básica. In *Docente*. Disponible in https://docente.ifrn.edu.br/felipepinto/disciplinas/2012.2/disciplinas/informatica-basica
26. Godwin-Jones, R. (2005). Messaging, gaming, peer-to-peer sharing: Language learning strategies & tools for the millennial generation. *Language Learning & Technology, 9*(1), 17–22.
27. Turnbull, M. L. (2002). The practical use of email lists as class discussion forums in an advanced course. In *International Conference on Computers in Education, 2002. Proceedings* (pp. 1418–1420).
28. Almeida, R. G. (2015). O aumento do engajamento no aprendizado através da gamificação do ensino. *Revista do Seminário Mídias & Educação.*
29. Rienzo, T., & Han, B. (2008). Microsoft or Google Web 2.0 tools for course management. *Journal of Information Systems Education, 20*(2).
30. Chang, E. (2005). *eHub Interviews Writely*. Disponible in https://web.archive.org/web/20110722190058/http://emilychang.com/ehub/app/ehub-interviews-writely/.
31. Hamburguer, E. (2013). Google Docs began as a hacked together experiment, says creator. In *The Verge*. Disponible in https://www.theverge.com/2013/7/3/4484000/sam-schillace-interview-google-docs-creator-box.
32. Riley-Huff, D. A. (2010). Using google wave and docs for group collaboration. *Library Hi Tech News.*
33. Srba, J. (2010). An experiment with using Google tools for project supervision at tertiary education. In *Proceedings of the 11th International Conference on Computer Systems and Technologies and Workshop for PhD Students in Computing on International Conference on Computer Systems and Technologies* (pp. 430–435).
34. Miseviciene, R., Budnikas, G., & Ambraziene, D. (2011). Application of cloud computing at ktu: Ms Live@Edu case. *Informatics in Education, 10*(2), 259–270.
35. Brodahl, C., Hadjerrouit, S., & Hansen, N. K. (2011) *Collaborative writing with Web 2.0 technologies: Education students' perceptions.*
36. Castellanos Sánchez, A., & Martínez de la Muela, A. (2013). Trabajo en equipo con Google Drive en la universidad online. *Innovación Educativa (México, DF), 13*(63), 75–94.
37. Boellstorff, T., Nardi, B., Pearce, C., & Taylor, T. L. (2013). Words with friends: Writing collaboratively online. *Interactions, 20*(5), 58–61.
38. Boellstorff, T., Nardi, B., Pearce, C., & Taylor, T. L. (2012). *Ethnography and virtual worlds: A handbook of method*. Princeton: Princeton University Press.
39. Naik, A. B., Ajay, A. K., & Kolhatkar, S. S. (2013). Applicability of cloud computing in academia. *Indian Journal of Computer Science and Engineering, 4*(1), 11–15.
40. Baiges, E. B., & Surroca, N. V. (2014). Valoración del uso de las herramientas colaborativas Wikispaces y Google Drive, en la educación superior. *Edutec. Revista Electrónica De Tecnología Educativa, 49,* a283–a283.
41. Olson, J. S., Wang, D., Olson, G. M., & Zhang, J. (2017). How people write together now: Beginning the investigation with advanced undergraduates in a project course. *ACM Transactions on Computer-Human Interaction (TOCHI), 24*(1), 1–40.
42. Wang, D. (2018), DocuViz. In *Chrome Web Store*. Disponible in https://chrome.google.com/webstore/detail/docuviz/hbpgkphoidndelcmmiihlmjnnogcnigi?hl=en.
43. Abrams, Z. I. (2019). Collaborative writing and text quality in Google Docs. *Language Learning & Technology, 23*(2), 22–42.
44. Storch, N. (2002). Patterns of interaction in ESL pair work. *Language Learning, 52*(1), 119–158.
45. Abrams, Z. (2016). Exploring collaboratively written L2 texts among first-year learners of German in Google Docs. *Computer Assisted Language Learning, 29*(8), 1259–1270.
46. Liu, H. K., Hung, M. J., Tse, L. H., & Saggau, D. (2020) Strengthening urban community governance through geographical information systems and participation: An evaluation of my Google Map and service coordination. *Australian Journal of Social Issues.*

47. Mohan, V., Kumar, C. G., Yuvaraj, J., Krishnan, A., Amarchand, R., & Prabu, R. (2020). Using global positioning system technology and Google My Maps in follow-up studies—An experience from influenza surveillance study, Chennai India. *Spatial and Spatio-Temporal Epidemiology, 32,* 100321.
48. Cilliers, L. (2017). Wiki acceptance by university students to improve collaboration in higher education. *Innovations in Education and Teaching International, 54*(5), 485–493.
49. Garcia, E., Moizer, J., Wilkins, S., & Haddoud, M. Y. (2019). Student learning in higher education through blogging in the classroom. *Computers & Education, 136,* 61–74.
50. Microsoft. (2020). *Skype makes it easy to stay in touch.* Disponible in https://www.skype.com/en/.
51. Akbaba, Y., & Başkan, F. (2017). How to merge courses via Skype™? Lessons from an International Blended Learning Project. *Research in Learning Technology, 25.*
52. Etherpad. (s.d.). Disponible in https://etherpad.org/.
53. Dropbox. (s.d.). Disponible in https://www.dropbox.com/.
54. Hunsinger, D. S., & Corley, J. K. (2012). An examination of the factors influencing student usage of dropbox, a file hosting service. In *Proceedings of the conference on information systems applied research* (Vol. 2167, p. 1508).
55. Google. (2020). Spark learning with G suite for education. In *For Education.* Disponible in https://edu.google.com/products/gsuite-for-education/?modal_active=none
56. Microsoft. (2020). Office 365 Education. In *Microsoft Education.* Disponible in https://www.microsoft.com/en-us/education/products/office.
57. Robillard, P. N., & Robillard, M. P. (2000). Types of collaborative work in software engineering. *Journal of Systems and Software, 53*(3), 219–224.
58. Steek, R. (2007). Your secured online space. In *Internet Archive.* Disponible in https://web.archive.org/web/20071012005211/http://www.steekr.com/index.php?m=d6afab95&a=e2c90ed8.
59. F-Secure. (2012). Online Backup. In *Internet Archive.* Disponible in https://web.archive.org/web/20120611144121/http://www.f-secure.com/fr/web/home_fr/backup/online-backup/overview.
60. Steek, R. (October 2007). UWA. In *Internet Archive.* Disponible in https://web.archive.org/web/20071011032706/http://blog.steek.com/steekr/index.html.
61. F-Secure. (2011). Fermeture du service SteekR. In *Internet Archive.* Disponible in https://web.archive.org/web/20120512222629/http://www.f-secure.com/fr/web/home_fr/support/support-news/view/story/528914
62. Uploadingit. (2008) Share your files with the world. In *Internet Archive.* Disponible in https://web.archive.org/web/20080923141229/http://uploadingit.com/home.
63. Pichai, S. (2012) Introducing Google Drive... yes, really. In *Google Official Blog.* Disponible in https://googleblog.blogspot.com/2012/04/introducing-google-drive-yes-really.html.
64. Mazzon, J. (2006). Writely so. In *Official Blog.* Disponible in https://googleblog.blogspot.com/2006/03/writely-so.html.
65. Microsoft. (2007). Microsoft acquires FolderShare, a file-synchronization technology provider. In *Internet Archive.* Disponible in https://web.archive.org/web/20070705184347/https://www.foldershare.com/info/company/aboutUs.php.
66. SC. (2007) Storage in the sky. In *Windows live team blog archive.* Disponible in https://windowslivearchive.wordpress.com/2007/08/10/storage-in-the-sky/.
67. Jones, C. (2007) Test drive the new windows live suite. In *Windows Live team blog archive.* Disponible in https://windowslivearchive.wordpress.com/2007/09/05/test-drive-the-new-windows-live-suite/.
68. BK. (2007). Store more with SkyDrive. In *Windows live team blog archive.* Disponible in https://windowslivearchive.wordpress.com/2007/10/11/store-more-with-skydrive/.
69. Antonia. (2007). Bigger, better, faster SkyDrive!. In *Windows Live team blog archive.* Disponible in https://windowslivearchive.wordpress.com/2008/02/21/bigger-better-faster-skydrive/.

70. The FolderShare Team. (2007). FolderShare—new beta, new blog!. In *Windows live team blog archive*. Disponible https://windowslivearchive.wordpress.com/2008/03/10/foldershare-new-beta-new-blog/.
71. Choney, S. (2014). Microsoft announces SkyDrive will soon be renamed OneDrive. In *Official Microsoft Blog*. Disponible in https://blogs.microsoft.com/blog/2014/01/27/microsoft-announces-skydrive-will-soon-be-renamed-onedrive/.
72. Gavin, R. (2014). OneDrive for everything in your life. In *Microsoft 365*. Disponible in https://www.microsoft.com/en-us/microsoft-365/blog/2014/01/27/onedrive-for-everything-your-life/.
73. Google. (2020). *Access your files and sync them anywhere*. Disponible in https://www.google.com/drive/download/.
74. Apple Inc. (2020). Google Drive. In *App Store Preview*. Disponible in https://apps.apple.com/us/app/google-drive/id507874739.
75. Apple Inc. (2020). Microsoft OneDrive. In *App Store Preview*. Disponible in https://apps.apple.com/pt/app/onedrive/id823766827?mt=12.
76. Google. (2020). Google Drive. In *Google Play*. Disponible in https://play.google.com/store/apps/details?id=com.google.android.apps.docs&hl=en.
77. Microsoft. (2020). *OneDrive*. Disponible in https://onedrive.live.com/about/en-za/download/.
78. Google. (2020). Google Drive. In *Chrome Web Store*. Disponible in https://chrome.google.com/webstore/detail/google-drive/apdfllckaahabafndbhieahigkjlhalf?hl=en-US.
79. Google. (2016). OneDrive. In *Chrome web store*. Disponible in https://chrome.google.com/webstore/detail/onedrive/nffchahhjecejoiigmnhhicpoabngedk?hl=en.
80. Docsecrets. (2018). *Home*. Disponible in https://www.docsecrets.net/.
81. Google. (2020). DocSecrets. In *G Suite Marketplace*. Disponible in https://gsuite.google.com/marketplace/app/docsecrets/933719089841?pann=cwsdp&hl=en.
82. IFTTT. (s.d.). *OneDrive*. Disponible in https://ifttt.com/onedrive.
83. Google. (s.d.). *Gmail*. Disponible in https://www.google.com/gmail/about/#.
84. Google. (s.d.). *Google Photos*. Disponible in https://www.google.com/photos/about/.
85. Google. (s.d.). Sites. In *Google Cloud*. Disponible in https://gsuite.google.com/intl/en/products/sites/.
86. Google. (2020). Clear Google drive space & increase storage. In *Google drive help*. Disponible in https://support.google.com/drive/answer/6374270?hl=en.
87. Microsoft. (2020). Plans. In *OneDrive*. Disponible in https://onedrive.live.com/about/en-US/plans/.
88. Google. (2020). Work with microsoft office files. In *G Suite Learning Center*. Disponible in https://support.google.com/a/users/answer/9308757?hl=en.
89. Microsoft. (2020). This is your 365. In *Office*. Disponible in https://products.office.com/en-us/home?omkt=en-US&rtc=1.
90. Apache. (2020). Apache OpenOffice 4.1.7 released. In *Apache OpenOffice*. Disponible in https://www.openoffice.org/.
91. Google. (2020) Share and collaborate in my drive. In *G Suite Learning Center*. Disponible in https://support.google.com/a/users/answer/9310248?hl=en.
92. LibreOffice. (2020). Disponible in https://www.libreoffice.org/.
93. Microsoft. (2020). Learn about file formats. In *Office*. Disponible in https://support.office.com/en-us/article/Learn-about-file-formats-56DC3B55-7681-402E-A727-C59FA0884B30.
94. Google. (2020). Microsoft OneDrive. In *Google Play*. Disponible in https://play.google.com/store/apps/details?id=com.microsoft.skydrive&hl=en.
95. Google. (2020). Make Google Docs, Sheets, Slides & Forms public. In *Docs editors help*. Disponible in https://support.google.com/docs/answer/183965?hl=en&co=GENIE.Platform=Desktop.
96. Microsoft. (2020). Embed files directly into your website or blog. In *Office support*. Disponible in https://support.microsoft.com/en-us/office/embed-files-directly-into-your-website-or-blog-ed07dd52-8bdb-431d-96a5-cbe8a80b7418

97. Microsoft. (2020). Share OneDrive files and folders. In *Office*. Disponible in https://support.office.com/en-us/article/share-onedrive-files-and-folders-9fcc2f7d-de0c-4cec-93b0-a82024800c07.
98. Rene Jackson 203. (April 2019). Formatting Issues when uploading Word docs in .docx format. [Post forum]. In *Docs Editors Helps*. Disponible in https://support.google.com/docs/thread/4313531?hl=en.
99. Sarton85. (November 2011). Why is the layout of text in Google Documents inconsistent between browsers? And what to do about it?. [Post forum]. In *Docs Editors Helps*. Disponible in https://support.google.com/docs/forum/AAAABuH1jm0dqRzrdJMBM8/?hl=en.
100. Leme, C. (Febrary 2016). Word 2013 perde formatação em outras versões. [Post Forum]. In *Community*. Disponible in https://answers.microsoft.com/pt-br/msoffice/forum/all/word-2013-perde-formata%C3%A7%C3%A3o-em-outras/44c3e0fe-78c2-4848-9426-04fa1e10f5bb.
101. Ahmed-Nacer, M., Urso, P., Balegas, V., & Preguiça, N. (October, 2013). Concurrency control and awareness support for multi-synchronous collaborative editing. In *9th IEEE International Conference on Collaborative Computing: Networking, Applications and Worksharing* (pp. 148–157). IEEE.
102. Denzin, N. K. (1989). *Interpretive biography* (Vol. 17). Sage.
103. Hatch, J. A. (1995). Life history and narrative: Questions, issues, and exemplary works. *Life History and Narrative, 1*, 113.
104. Talja, S. (2002). Information sharing in academic communities: Types and levels of collaboration in information seeking and use. *New Review of Information Behavior Research, 3*(1), 143–159.
105. Maio, V., Campos, F., Monteiro, M. E., & Horta, M. J. (2008) Com os outros aprendemos, descobrimos e... construímos-um projecto colaborativo na plataforma Moodle. *Educação, Formação & Tecnologias, 1*(2), 21–31, ISSN 1646-933X.
106. Korbel, J. O., & Stegle, O. (2020). Effects of the COVID-19 pandemic on life scientists. *Genome Biology, 21*, 113.
107. Kawasaki, G. (1990). *MacIntosh way: The art of Guerrilla Management*. HarperTrade.
108. Microsoft. (2020). Evangelism. In *Careers*. Disponible in https://careers.microsoft.com/professionals/us/en/c-evangelism.
109. Zotero. (2010). Zotero can be used with Google Docs in the same way as with plain text documents or emails... . In *Documentation*. Disponible in: https://www.zotero.org/support/google_docs?rev=1282206336.
110. Luan, A., Momeni, A., Lee, G. K., & Galvez, M. G. (2015) Cloud-based applications for organizing and reviewing plastic surgery content. *Eplasty, 15*.
111. Snowden, E. (2019) *Permanent record*. New York: Macmillan.
112. Google. (s.d.). My maps. In *Google Maps*. Disponible in https://www.google.com/intl/en/maps/about/mymaps/.

Chapter 5
Good Practices for Online Extended Assessment in Project Management

C. Popescu and L. Avram

Abstract This chapter aims to represent an element in the area of dissemination of project results and a guide of good practices in terms of project implementation. In the management of academic educational projects, a careful monitoring of the complex activities is necessary and it requires the implementation project team a permanent effort to verify the compliance with the planned level of the results, in different phases of the project. This monitoring includes the interrogation of the project participants regarding the organization of the project activities, the continuous updating of the content of the activities for providing adequate training services (both theoretical and practical), the level of satisfaction of the beneficiaries, etc. Therefore, if the analysis is related to the international projects and to the multicultural teams, where there are many partners and different heterogeneous target groups it is recommended to use a simple and friendly tool, with adequate intimacy level and with the possibility of obtaining immediate information or useful feedback (generated by the software capabilities). So, we could use the features of the free tool owned by Google, named Google forms, and design an adequate questionnaire/survey.

Keywords Assessment tool · Online · Feedback · Questionnaires · Educational project · Multicultural teams · Target group · Good practices · Satisfaction level · Compliance level · Automatic report · Google forms

---

C. Popescu (✉)
Business Administration Department, Petroleum-Gas University of Ploiesti, 39 Blvd. Bucureşti, Ploiesti, Romania
e-mail: cpopescu@upg-ploiesti.ro

L. Avram
Department of Well Drilling, Extraction and Transport of Hydrocarbons, Petroleum-Gas University of Ploiesti, 39 Blvd. Bucuresti, Ploiesti, Romania
e-mail: avram_lazar55@yahoo.com

R. Babo et al. (eds.), *Workgroups eAssessment: Planning, Implementing and Analysing Frameworks*, Intelligent Systems Reference Library 199,
https://doi.org/10.1007/978-981-15-9908-8_5

5.1 Introduction

Successful implementation of educational projects takes into account how the activities included in the project are carried out. The sustainability of such a project is given by the way in which the participants in the project, that is, the members of the target groups, benefit from proper support regarding the delivery of knowledge and the acquisition of skills. Therefore, the project teams must ensure that they will achieve the best results, and this can be verified if, after the activities unfold, they assess the level of satisfaction of those who have benefited from the appropriate development of these activities.

Basically, the current project, described in this chapter, had as a main goal the designing and implementation of a Master's programme in the field of oil and gas engineering within some universities from Lebanon (please see the details from the subchapter Project data) through a joint partnership provided by the academic staff and experts from partner universities of Europe.

The complexity of the projects that propose the implementation of Bachelor or Master's degree programs compels to the use of objective tools, simple to use and which automatically generate a global interpretation of the feedback that participants and, at the same time, beneficiaries of specific activities can deliver to those who are directly involved in the development of the respective projects.

The mechanism used in these situations must have certain characteristics such as suppleness, flexibility, simplicity, assurance of anonymity, transparency of the information delivered, and automatic creation of objective reports, which should obviously help to permanently improve the results of the implementation of these programs of study. Under these circumstances, one wants to find a simple and useful way to answer all these requirements. Obviously, the initial task envisaged was to integrate different types of questionnaires into the feedback analysis. Why this? The answer is simple. The implementation of a program of study involves various and complex types of educational and research activities.

The chapter aims to promote good practices in strategic academic partnerships in the field of training and professional development in higher education. The project implementation also involved the application of quality management principles to ensure an efficient transfer of knowledge and build new skills. The project developed in an international academic partnership refers to the creation and development of a curriculum for a master's program in the field of Gas and Oil Processing. The trainers used in the work package (WP) associated with formal training (called *Development*) come from three European universities, while the beneficiaries of the project belong to a number of four universities in Lebanon.

In addition, the work package entitled *Quality plan* requests, among other objectives, to check the quality level and compliance percentage in relation to the accomplishment of the main scheduled activities of the project (Table 5.1).

The problems considered both in the project implementation and in the developments presented in this chapter refer to:

Table 5.1 Project activities

Scheduled activity	Explanations
Workshop (W)	"*training for trainers*" (for the teachers from Lebanon universities, that will teach the students participating in the master programme)
Webinar (Web)	Presentations and thematic discussions carried out with the help of the online technology
Internship (I)	Training of participating students in the master's degree program, in the sense of involving students in practical activities specific to the oil and gas field (within companies from the oil and gas industry or in laboratories belonging to European universities involved in the project)
Master thesis (MT)	Completing the thesis according to the curriculum and in relation with the internships content

- the heterogeneity of the academic partnership based on a number of nine partners coming from four countries has led to the creation of a multicultural environment with various specificities that must be known and managed properly;
- the relatively large distances between the universities participating in the project (from universities located in Western Europe, to a university located in Eastern Europe and reaching universities belonging to a Mediterranean country in Southwest Asia, more precisely Lebanon);
- the need to generate a unitary structure in terms of how workshops and internships take place;
- designing efficient and flexible procedures regarding the implementation of the project activities through which to ensure a fast and objective feedback related to the level and content of the activities so that any shortcomings or dissatisfaction signalled by the participants in the project activities can be corrected in due time;
- the integration of a master's degree programme that corresponds to the academic requirements at an international level but also to the realities of the economic environment in Lebanon.

In the following paragraphs, the main sequences and relevant issues in the international project implementation will be described accordingly. This description part will help the readers form a better understanding of the project content and development.

In the first stage, the curriculum of the program of study should be designed based on the expertise of those who will teach courses to future instructors. The program must respond to specific requirements such as correspondence with other similar programs at the international level and ensuring the defining elements at the theoretical and practical levels. Therefore, first checking of accuracy is made based on feedback obtained from specialists recognized in the academic environment for their professional achievements, but also in relation to the opinions expressed by some practitioners in the chosen field. Then it is necessary to train the instructors who will teach future undergraduate or Master students. In this case, according to the established curriculum, the subjects to be included in the materials and presentations of the experts who will teach the future trainers are determined for each discipline. These new instructors are chosen by each beneficiary entity (each Lebanese university) based on their basic training and in relation to the requests that these universities have received from potential attendees of these programs of study. The future trainers have had useful interventions during the presentations made during the workshops, so that the experts can review the materials in an appropriate way. In addition, during the implementation of such projects, as they are international partnerships, it is useful and even recommended to organize webinars on specific and up-to-date topics. In this case, the participants to discussions intervened with constructive ideas and suggestions which helped to outline useful materials for both the future instructors and the future undergraduate or Master's degree students.

In carrying out the activity of implementing the new program of study, a first integration of the plan of study can be included, as an experimental activity, after the necessary accreditations have been obtained, in accordance with the legislation in force. In this respect, each entity that will implement the new program of study may experience the training of a first series of undergraduate or Master's Degree students who will participate in carrying out this program of study as a premiere. This program will benefit from the support of the new instructors trained within the project and under the careful supervision of those who presented the defining materials of the curriculum during the workshops. Like any participant in a program of study, the beneficiaries have to design a final scientific project (called a bachelor's thesis or a dissertation thesis). Even this kind of activities, once carried out, requires a mandatory feedback from the students, which will facilitate in a timely manner the adjustment of all that must be obtained as a concrete result by each graduate of the program of study. An important component of such an educational project also aims to include practical activities that will help the future graduate to better understand the elements specific to the field in discussion. This type of activity refers to the participation in internships organized by profile companies. At the end of these activities, the beneficiaries provide feedback to the employers in order to adapt a program of activities oriented towards the students' requirements.

This range of activities must be designed very carefully, then it is necessary to find the appropriate means for putting into practice each activity, so that finally the control is carried out by conducting surveys based on customer satisfaction questionnaires [1] (given to the members of the target group). In fact, for each project activity, different target groups were set up based on what was previously mentioned in this subchapter in connection with the WP *Quality Plan* and its objectives. Later the paragraphs dedicated to the case study will present details about these target groups.

At the same time, the relevance of the title and content of the chapter is related to the content of the WP called *Development* which has the following defining activities: *workshops* developed for future trainers in the Master's programme, *webinars* with topics dedicated to specialists in the field and *internships* for students who are integrated into the Master's programme and which require attendance to practical activities (within specialized companies or dedicated laboratories) in order to obtain and use the information which is required to complete the dissertation thesis. The complexity of this WP claims a careful monitoring by its coupling with WP *Quality*

Plan through which the quality of the activities carried out and the level of compliance with the project' objectives and outcomes are verified by means of questionnaires. One of the main goals of this particular project was to design a standard procedure in order to help improving the educational and practical training services delivered for any new Master programme. This procedure could be run iteratively and what was obtained through the participants feedback was not only to test the satisfaction level of the participants involved in different project' activities but also to create useful and effective information that can represent good practices for projects with similar content.

Another issue to be mentioned is the obligation to verify the level of compliance with the scheduled values of what represents project outcomes. And this goal can be achieved using appropriate means of monitoring and objective and rapid analysis of the feedback of the people involved in the project, both as implementers and as beneficiaries.

In relation to all those mentioned, the solution that can be taken into consideration is the use of online questionnaires.

At the same time, this chapter presents itself as a textbook of good practices giving examples of methods to overcome an important number of challenges that the project team faced during the project implementation: an academic network with a large number of partners, increased geographical representation (Western Europe, Central and South-Eastern Europe, South-West Asia), cultural diversity, different typologies regarding higher education systems.

The main question for this particular chapter is related to whether the online questionnaires, designed by using Google forms, are reliable to be taken into account in order to get full information about the project implementation and the achieved outcome. In fact, the main challenge in this research was to check the level of compliance of achieved project results with the scheduled values and the satisfaction level of the participants in different kinds of activities carried out within the project: workshops, webinars, internships, etc. In this sense, the following paragraphs are mentioning some useful information related to the *means* that were designed and used especially to carry out the surveys necessary to support the WP *Quality Plan* aims.

Only one of the survey pillars was to verify the level of satisfaction of the beneficiaries of theoretical and practical training. The other survey pillars referred to multiple aspects such as: organization of activities, actual content/proposed agenda for each activity, methods and tools used in training, efficiency project implementation process etc. The surveys were designed in relation to the structure and content of the issues in the questionnaires (see for example the four-section structure of the questions in the questionnaire dedicated to the workshops).

Consequently, analyzing the satisfaction of beneficiaries of different products and services (including the educational domain), various quantitative and/or qualitative tools can be used. The most important tool used by the surveyors, for asking questions and recording the answers, is the questionnaire [2]. The starting point in designing any questionnaire is the determination of the type and volume of information to be collected in such a way [3]. For this, the *purpose* of the marketing research in which the questionnaire will be involved must be clearly defined [4]. The core of any questionnaire consists of the questions it contains. In writing the questions, the starting point cannot be other than *the information needed* to solve the problems required by the marketing and sociological research that makes it necessary to use questionnaires [5, 6]. The drafting of the questions to be included in a questionnaire, at the same time, concerns both the science and the art of communication [7, 8]. The questionnaires aim to obtain information impossible (or much more difficult) to obtain otherwise, with the help of which researchers try to solve some concrete problems targeted by the marketing research that involves them [9]. In these days, when ICT (Information and Communication Technologies) are used in many sectors, including education, there are various tools that address various topics: innovative learning software; adaptive learning applications; online courses; virtual reality in education; social media learning; digital assessment and test [10]. These tools are designed to support accordingly different activities in teaching, learning, and assessment.

5.2 Writing and Using an Online Questionnaire

The advantage of using a questionnaire that is completed online [11] is that the answers to the questionnaire are recorded directly in a database (so you do not have to enter the questionnaires manually), the time required to complete the questionnaire is shorter (because the person answering it just clicks on the variant), you have greater control over the answers (there is the possibility of warning the respondent who skipped a question) and you can distribute the questionnaire more easily to a larger number of people (just send the link to the questionnaire). Of course, there are disadvantages such as the lower response rate (an online questionnaire is more impersonal, and if the person does not know you they may not respond), it is difficult to apply the questionnaires at the same time (not everyone completes the questionnaire at the same time), you cannot control the time allocated to the questionnaire by each person and the respondent needs to have access to a computer and to the internet connection (which means you can address only a population with certain characteristics and with a certain standard of living).

The following paragraphs will demonstrate why Google Forms has important advantages that confirm many of the features required by feedback requests [12, 13], in order to question the satisfaction of some beneficiaries of educational services: suppleness, flexibility, simplicity, ensuring anonymity, transparency of information delivered, automatic creation of objective reports, etc. Also, through the Google forms, the questionnaires could offer info that cover many other issues related to the educational project implementation.

In this regard, Google contains a diverse set of apps: from documents to spreadsheets to operations in cloud [14]. In this sense, we mention some online collaboration tools: Google Docs, Google Sheets, Google Slides, Google Forms, which facilitate teaching and learning activities [15]. More recently, there is a practice, applied especially in the academic area, through which it is desired to collect useful information about course instruction using the feedback generated by completing the questionnaires using Google forms [16]. There are countless recommendations regarding the use of questions, their creation or the exploitation of graphics and data through Google forms [17]. Google Sheets are used for well-defined purposes in order to protect data, validate data configuration, or design and edit charts on sheets [18].

Over time many platforms that allow the creation of questionnaires, surveys, or online tests have been developed, each of these platforms demonstrates their strengths in order to meet the needs of their users [13, 19]. Among them, Google Forms is the one that best suits the business goals and not just that. Google Forms allows the creation of questionnaires which could be sent later and the results can be analyzed in real-time [40]. Surveyors can interpret the data with the help of the centralization done by Google Forms or process it with Google Sheets. Being extremely versatile, this tool allows the creation of questionnaires of all types and on various topics. The results collected through Google Forms are automatically recorded on an Excel spreadsheet, giving the author of the questionnaire or survey the possibility that these results be verified in real-time throughout their performance [41]. Google questionnaires automatically store answers in the form. Each answer is saved in "answers" (the tab at the top of the form) and is updated in real-time as users answer the questions.

In terms of assessment, the types of assessment commonly used to evaluate skills in the educational field refer to formative assessment, summative assessment and diagnostic assessment [20].

The mentioned questionnaires will include, among others, a suite of questions (a section) that refers to the Diagnostic Assessment method. This particular method aims to establish in advance the knowledge level of the student at the beginning of the training or practice [21]. This check is also necessary to establish what is the real benefit generated by participating in the training activity or in specialized practice.

5.3 State of Art

There are not many papers that describe good practices or that contain useful ideas regarding the design and the use of online assessment systems for the activities carried out within an educational project in order to get, in due time, desired outcomes. In addition, there are not many guidelines on effective practices in designing processes for assessing training activities using online capabilities. However, some contributions will be presented in the following paragraphs. For instance, there are designed online systems that facilitate evaluation and feedback created for the learning processes monitoring and for successfully implementing educational projects [22]. These systems can help to improve learning processes and achieve desired outcomes during the implementation of educational projects.

A critical issue pursued in the implementation of the project described in the case study has in view, the correlation of the principles enunciated in WP *Quality Plan* with the necessary aspects in checking the proper evolution of each activity in the project. In this regard, there are approaches that propose a complex assessment framework that verifies the resources used in the training activities carried out through educational projects [23]. Additionally, there are some contributions related to the creation of interactive assessment models (i.e. online type) that make full use of the ICT features and facilities [24].

Online assessment tools can be used to analyze the correlation between Project Management principles and practices, and project results obtained during implementation [25].

Another important challenge concerns those techniques and tools that have an essential contribution to the success of a project [26]. This kind of analysis wants to demonstrate the potential and the contribution offered by each tool to the improvement of the performances during the daily implementation of the project. As we will show in the case study, a valuable tool in the successful conduct of educational projects is the flexible construction and use of the online questionnaires through Google forms. This will allow the adaptation and immediate correction of the shortcomings found in the theoretical and practical training activities, issues related to the organization of activities, the teaching process, the content of materials and presentations, the trainers, the results to be obtained, the usefulness of the activity in relation to the goals pursued by the project, etc.

At the same time, the entities that organize training processes that are intended to be efficient are concerned with the design and use of the online assessment systems that have in view the continuous verification of the teaching process improvement and the provision of new skills [27, 28].

The assessment of academic educational processes is viewed by specialists differently. There are still supporters of the traditional university assessment (looking to check the knowledge level), but more and more are advocating for the so-called *alternative assessment* in which the evaluation is designed in an integrative way, which includes both training processes (classical, online or hybrid) and monitoring the real performance generated by the participants, up to a certain point from the training period [29]. The online environment offers important advantages such as adaptability and flexibility compared to traditional assessments. This change of modality, which is in fact related to the assessment tools used, is due to the evolution of ICT type technology because, in this new scenario, complex quantitative–qualitative evaluations are allowed [30].

There are more and more opinions regarding the use of ICT in terms of the quality of the learning processes and which help to generate effective assessment environments as it is possible: interactivity, flexibility, student/trainee centeredness, etc. [31, 32]. It is also worth noting that there are researches that confirm the need to use alternative forms of assessment to allow useful feedback on how training processes are conducted [33]. This idea reinforces the need to use ICT-based tools in connection with the assessment of the development of educational projects aimed at complex processes of theoretical and practical training, such as the introduction and development of a Master's study program.

5.4 Case Study: The European Erasmus + GOPELC Project

5.4.1 Project Data

This case study considers the presentation of some experiences as good practices, by carrying out and monitoring the daily project activities, related to the analysis of the satisfaction of the multiple beneficiaries of the project (teachers participating in training, experts from the economic environment, students integrated into internships, etc.) and to the analysis of the compliance level of the results generated during the project implementation to the proposed project outcomes [34]. An example described briefly in Introduction which we consider relevant for the topic proposed in this chapter, refers to an Erasmus + project (project funded by EU funds within the framework of the Erasmus + Program) developed between 2016 and 2018 and for which an academic consortium consisting of 7 universities was constituted: 4 from Lebanon, one from Romania, one from France and one from Sweden. The consortium has generated an academic network made up of well-known universities in the fields of oil drilling and extraction, transport of hydrocarbons, oil, and natural gas processing, engineering for environmental protection and chemical engineering (the main topics of interest of this project). More precisely the consortium consists of the following universities: University of Petroleum-Gas in Ploiesti, Romania (project coordinator); Ecole de Mines of Saint Etienne, France; Royal Institute of Technology, Stockholm, Sweden; Lebanese University, Beirut, Lebanon; Notre Dame University, Beirut, Lebanon; University of Balamand, Beirut, Lebanon; Beirut Arab University, Lebanon. Overall, the project considered the design, development and implementation of a Master's degree program in the field of oil and gas engineering in 4 universities from Lebanon (one public and three private) based on the expertise provided by the specialists coming from partner universities in Europe.

This project was carried out in perfect agreement with the European Commission's actions aiming to deepen the knowledge in the fields of extraction technologies and unconventional methods to minimize potential risks on health and on the environment through academic partnership. The partnership aimed to bring together practitioners from industry, research, academia and civil society, so as to ensure a fair and balanced exchange of ideas.

The following few paragraphs are dedicated to describing the main types of activities included in the project. The *workshop* aimed at giving lectures on well-defined topics and in relation to a curriculum consisting of 15 disciplines. Each discipline had a number of topics mentioned in the curriculum, and the trainer appointed to give the presentation had to organize his/her lecture covering 60% of the topics in their presentation, and offering 100% of the topic through materials. The instructors were represented by academic staff and experts in the fields of the existing disciplines in the curriculum and belonged to the European universities within the consortium created for the project. The target group was represented by academic staff from each beneficiary university in Lebanon and who will subsequently be trainers for the students involved in the Master's program.

The webinar (using Skype sessions) was designed to be a forum for extensive discussion and exchange of ideas between practitioners, experts in the field and academic staff on specific topics, niche issues and current affairs in the field of oil and gas engineering. Workshop trainers and the beneficiaries of these workshops participated as well in these webinars.

The internship was designed as a practical training course (carried out either within oil and gas companies from Europe or in the specialized laboratories of European universities belonging to the project consortium) The beneficiaries of the internship were students in the period of graduation (students in the last year) and who had to document themselves properly and with useful and up-to-date information for the completion of the final thesis.

The first year of the project was dedicated to the design, preparation of the accreditation of the new course programs and the development of training and teaching materials in detail. The design of the study program started from the analysis of the requirements of the Lebanese universities and from the examination of the contents existing in the Lebanese universities, the course programs being based on the Teaching and Learning Methodology. Staff members from EU countries have prepared detailed teaching and training materials in relation to the developed program. Five workshops were conducted for the training of the teaching staff in the partner countries and for a good transfer of experience and knowledge (3 in Romania, one in France and one in Sweden).

Two online seminars (webinars) were organized by the universities of France, and Sweden on specific and topical subjects in the oil and gas industry. In the second and third year, the foundations of the new Master's program were laid. The duration of the study program is 2 years, which is in perfect agreement with the Bologna requirements. This Master's program included basic courses such as professional English and specialized training activities for developing skills for the job market [35]. For this purpose, for two consecutive years (2017 and 2018, in Romania, France and Sweden), internships of at least one month have been carried out in laboratories at EU universities and at oil and gas companies (so-called specific practical stages). The program ends naturally with the elaboration of the dissertation paper.

The participants in all these project activities were in sufficient numbers so that the surveys conducted through the questionnaires were representative in relation to the project objectives. Thus, in the case of the workshops there were a number of 41 people acting as trainees (teachers belonging to the beneficiary universities from Lebanon trained by 26 academic staff and experts coming from EU partners), while the webinars were attended by a total of 27 people (teachers, practitioners and experts in the oil and gas industry; 13 people attended the first webinar and 14 people were present in the second webinar). Moreover, in the case of the internships organized in Europe, 27 students belonging to Lebanese partner universities were involved, while the number of people that completed the dissertation thesis within the project was 40 (all being students from Lebanon).

From the point of view of carrying out educational activities, an approach based on learning outcomes was taken into account. A modular course model was adopted. Members of the Lebanon partner universities participated in the workshops at European universities, being able to develop their own teaching material based on their own education system, on previous student experience and on the needs of professional organizations.

The methodologies have been thoroughly tested in the EU Member States and participating universities. But scientific, technological and operational methodologies and practices were not simply transferred. In the preparation phase of the Master's program, the experiences and practices developed in the EU were discussed and shared with Lebanese partners.

The approach based on the learning outcomes was used for the development of the discipline programs. This approach emphasizes the skills and competences that will be acquired by the student through the learning process. Teachers from EU universities have developed teaching materials in English because it is the most commonly used language in the Lebanese academic community. Teaching materials included PowerPoint presentations, class notes and supporting material (videos, standards, guidance, papers and technical documents).

The workshops had an average duration of one week, 5 working days, 8 h a day. During the workshops, the teachers from the European universities presented the teaching material along with case studies, exercises and supporting material. A better interaction between European and Lebanese teachers has been established and it will continue in the coming years. The workshop was not only intended to explain the learning material and to transmit the scientific experience within the EU to Lebanese teachers, but also to discuss teaching methods and learning outcomes. During the workshops, technical visits were scheduled in the European laboratories and facilities here, but companies with oil and gas profiles were also visited. Last but not least, the main beneficiaries of the project, i.e. the universities in Lebanon, have had the opportunity to equip themselves with specific equipment for the laboratories that will serve these study programs and not only: simulators, chromatographs etc. When the questionnaire for the workshops was designed for each section(s) considered, we into account the consideration of different evaluation methods, based on different objectives in terms of assessment, as follow (Table 5.2).

The work packages related to this project were: WP1 Preparation, WP2 Development, WP3 Quality plan, WP4 Dissemination and WP5 Management. The following will describe aspects related to the WP3 Quality plan.

The purpose of the Quality Assurance Plan was to ensure the quality of the GOPELC project and the achievement of the specific objectives, targets and plans. The entity in charge with implementing this WP was KTH from Stockholm, Sweden. One of the main principles stated in the Quality plan claim that the general management of the project should follow the needs and satisfaction of target audience and relevant stakeholders should be at the top of the list while developing all project activities [36]. For carrying out a full and effective monitoring work the main tool used was Evaluation Questionnaires for each project event. Therefore, for each main activity, Prof. Joydeep Dutta and Mr. Bassam Kayal from KTH, were in charge of

Table 5.2 Workshop questionnaire sections in relation with evaluation method

Questionnaire section	Explanations/evaluation method
S1	The objective pursued was that of diagnostic type evaluation for which, among other things, the goal was to identify the level of knowledge and skills existing at the time of the training targeted by the topic of the workshop
S2	The objective was to obtain a relevant feedback from the students, based on which to identify in due time the trainer's gaps in teaching, in communication and problems regarding the debate of the topics in the agenda. The method chosen in this case was formative assessment
S3 and S4	Critical for the success of the workshops, aimed at the use of the Summative assessment method, as their purpose was to establish at what level of fulfilment the expected results were obtained. In this sense, the questions formulated had the purpose to verify the efficiency of learning at the time of the workshop, but also to obtain relevant reactions regarding the training itself and information about the benefits for the participants

designing specific questionnaires. In order to cover all the requirements related to the project complexity and implementation they proposed different questionnaires, by using Google Forms, and the next pages it will enlarge upon some of them. The Romanian team distributed these questionnaires among all the categories of people involved in the project. The access to the questionnaires can be made based on a request addressed to the board of the project consortium.

5.4.2 Methodology

The methodological approach considered the use of questionnaires. In this sense, the closed questions from the questionnaires, analyzed by quantitative methods, have generated firstly worksheets and then percentages, pie charts, tables and graphs. The qualitative approach generated by the answers obtained in relation to open question created critical analyzes without the use of numbers or calculations. The methodology considered, using online questionnaires, also wanted to respond to specific requirements: increased data collection speed, low-cost requirements and high levels of objectivity. This last requirement was obtained by the fact that the answers were confidential, and the processing and interpretation of the information generated was performed by a small team of 4 people (involved in managing the work package 3 in the project, entitled *Quality Plan*). In addition, in the design of the research, the closed questions from the questionnaires were associated to four aspects (Table 5.3).

In processing the feedback generated by the answers to the closed questions, they aimed to obtain individual scores (through a Likert scale from 0 to 4, i.e. 0 total unsatisfactory—minimum value and 4 total satisfactory—maximum value) for each

Table 5.3 Issues covered within closed questions

No.	Issues covered through the questionnaire sections
1	Organizing the activity and the quality of the information received by the workshop participants
2	The trainer/person who managed the activity (through the presentation and communication skills, qualification level, willingness to answer to the questions at any time)
3	Satisfaction level (in terms of organization, useful information, trainers skills, activities outcomes related to the project objectives)
4	Generated results and impact

Table 5.4 Questions grouping into the sections/clusters

Section/cluster	Content
S1	Organizing the activity and the quality of the information—questions 1, 2, 5, 6, 9 and 11
S2	Trainer skills—questions 4, 7, 8, 10 and 13
S3	Satisfaction level—questions 3, 12, 14, 15 and 19
S4	Generated results and impact—questions 16, 17, 18 and 20

question, an overall average score for each activity sustained and average scores associated with the four aspects mentioned above.

For the final processing, the groups of questions were integrated as follows (Table 5.4), in four sections (clusters).

The target group of the project refers to the participants in the major activities of the project: workshops, webinars and internships. It is important to mention that all four universities in Lebanon, partners in the project consortium, had representatives (teachers and/or students) in all these activities and most of them answered the questionnaires launched by the Project Management Board. The questions of the analysis conducted through questionnaires aim to identify the level of satisfaction of the beneficiaries of the activities carried out in the project and the level of compliance of the training service providers to the objectives initially established in the project proposal and adapted in the first part of the project implementation in relation to the realities on the ground. Respondents were asked to complete a computer questionnaire, received by email.

The advantages of computer questionnaires include their low price (virtually non-existent), the use of time more efficiently, while respondents do not feel pressured, and as a result they can respond when they have time, providing more accurate answers. Anyway, the main disadvantage of the computer questionnaires is that the respondents may not bother to answer and they could ignore to fill out the questionnaire. Regarding the projected results, a minimum average score was set up for the satisfaction level of the participants in every type of planned activity (workshop, webinar, internship, etc.) and that minimum average was of 3 out of maximum 4.

Table 5.5 Number of respondents for each workshop (W)

No	Activity type	Number of responses
1	W1	7
2	W2	13
3	W3	15
4	W4	8
5	W5	5

Procedurally, these questionnaires are accessed by everyone who owns the link to Google Forms. This link is made available upon completion of the scheduled activity (workshop, webinar, internship, Project meeting board, etc.). At the time of filling in online, the responsible entity receives notification regarding the completion of the questionnaire (in this case the persons responsible from KTH Sweden).

5.4.3 Survey Results

1. Workshops assessment

An important first questionnaire was the one dedicated to the participants in the workshops. To have a better image related to the e-assessment activity, linked with the workshops, all the persons that are answered to the online survey are mentioned in Table 5.5. As in most survey-related cases, when it comes to answering questionnaires, not all workshop participants answered. In fact, the response rates for the workshops (W) were: 87% for W1, 86% for W2, 83% for W3, 80% for W4 and 58% for W5.

The questionnaire included 20 closed questions with multiple choice answers and one last open question, with the possibility of completing with some personal comments and suggestions [37]. The questions focused first on general issues, followed by questions related to the trainer, the training session itself and the benefits obtained by participating in the workshop.

Then, a report showing the feedback delivered for each question by the target group can be automatically generated. This report contains data on the persons who answered the questionnaire, the universities and the departments to which the respondents belong. In October 2017, a first quality report was made regarding the development of the programmed workshops by inserting the generated results automatically generated by the platform created on Google forms. The project management team then interpreted these results, generating measures to improve the processes carried out.

Compared to other researches based on questionnaires [38] which took into account in the analysis of the survey results, segmentation criteria such as gender, age etc., the processing of the questionnaires in this particular case had in view the construction of a global image, without criteria for segmenting respondents, in relation to the objectives of these surveys. On this occasion, various types of tables were created that cumulated the types of response (including percentages) for the 20 closed questions defining the questionnaire dedicated to the workshops.

The percentage values were translated into scores associated with an adapted scale for quantifying the answers (through a Likert scale with score 0 meaning totally unsatisfactory and score 4 meaning totally satisfactory). This operation is presented in Table 5.6:

In order to offer a first conclusion, the next table (Table 5.7) was designed (through a Likert scale with score 0 meaning totally unsatisfactory and score 4 meaning totally satisfactory):

Table 5.6 Average scores for each question versus each workshop (W)

Workshop	W1	W2	W3	W4	W5
Question number					
1	3.85	3.30	3.30	3.40	3.20
2	3.71	3.23	3.20	3.20	3.40
3	3.57	3.15	3.20	3.10	3.20
4	3.57	3.46	3.40	3.20	3.40
5	3.42	3.38	3.30	3.20	3.20
6	3.71	3.46	3.13	3.20	3.40
7	3.42	3.46	3.53	3.30	3.20
8	3.31	3.46	3.30	2.70	3
9	3.57	3.15	3.30	3.20	3.20
10	4	3.38	3.73	4	4
11	4	3	3.46	4	4
12	4	3	4	4	3.2
13	4	4	4	4	4
14	4	3.69	4	4	4
15	4	4	4	4	4
16	4	3.69	4	4	3.20
17	4	3.38	3.73	3	3.20
18	4	3.69	4	3.50	4
19	3.57	3.30	3.26	3.20	3.40
20	3.85	3.69	3.73	3.50	3.40

Table 5.7 Average scores for the workshops (W)

Workshop	W1	W2	W3	W4	W5
Item					
Average score	3.77	3.45	3.59	3.48	3.48

Table 5.8 Average scores for the workshops (W)

Workshop	W1	W2	W3	W4	W5
Average score per Section (out of a maximum of 4)					
Average score for S1	3.70	3.25	3.28	3.36	3.40
Average score for S2	3.66	3.55	3.58	3.43	3.52
Average score for S3	**3.80**	**3.42**	**3.68**	**3.65**	**3.56**
Average score for S4	3.95	3.61	3.85	3.50	3.45

One can state that the average scores for each workshop are quite high near to the maximum mark, which can be translated as a very good level regarding the participants' overall appreciation within the workshops. A short notice related to the first workshop is that the score is a little bit higher than the others because the time for its preparation was two months longer than the preparation periods for the other four workshops (according to the Project Gantt chart).

Another kind of feedback could be obtained by analyzing the average scores obtained in each section within each workshop. Therefore, the following Table 5.8 was created:

Basically, each section can be analyzed and interpreted. The section of interest is Sect. 5.3 (S3), which is dedicated to the satisfaction level (highlighted here by red border). This section was analyzed with priority because the project management team was especially interested in providing all the elements of training, learning and didactic materials for the future trainers in the Master's program. As it is easy to notice, the average scores are still high with a slightly increased value for the first workshop. By checking the comments received from the participants (notes inserted in the space for the open question) the explanations for these small differences were as follows (Table 5.9):

In order to combine different means in expressing, in another way (i.e. using percentages), the satisfaction level for the workshops, the next two figures (Figs. 5.1 and 5.2) represent *pie-charts* diagrams, that are generated through Google Forms. These two diagrams refer to the situations characterized by extreme values (highest appreciation and the opposite one). More comments will be found in the Discussion and Conclusion sections.

Table 5.9 Main comments of the workshops participants to the open question

Workshop	Comments
W1	The agenda included three technical visits: two in-situ visits (an oil field belonging to the OMV –Petrom SA, Asset VI Muntenia Central and a pumping station of CONPET SA) and one visit to some specialized laboratories from Petroleum-Gas University: the mechanical processing laboratory; the laboratory for complex mechanical testing of Oil Country Tubular Goods; the command and data acquisition laboratory; the drilling simulator PIPE SIM laboratory
W2 and W5	No technical visits were planned

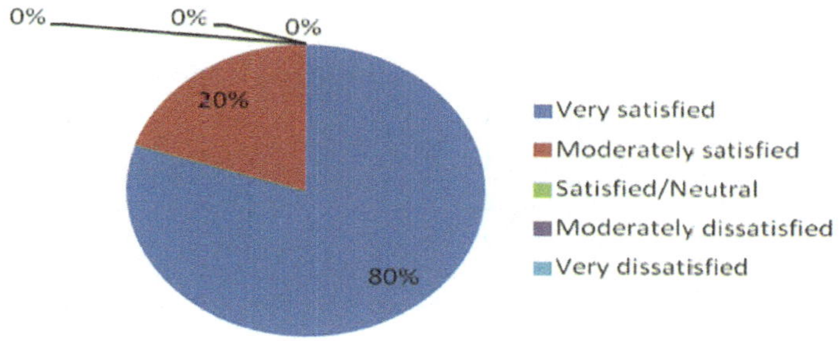

Fig. 5.1 Average percentage for S3 in the case of workshop 1

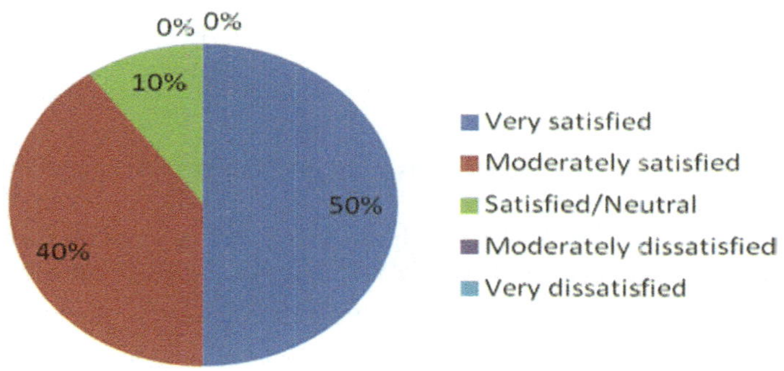

Fig. 5.2 Average percentage for S3 in the case of workshop 2

2. Webinars assessment

Moreover, another type of activity, scheduled in the project Gantt chart, was to carry out some webinars (in fact there were two: one organized by EMSE, France and the other one organized by KTH, Sweden). For these events questionnaires designed by KTH were set-up as well in order to get a proper feedback from the participants. For these questionnaires used Google Forms was used as well. The response rates for the webinars (Web) were: 54% for Web1 and 50% for Web2.

3. Internships assessment

Another important project outcome was the participation of the future master graduates in internships of at least one month, in order to get some practical skills and prepare the final thesis. These internships were carried out in laboratories at the EU universities and within oil and gas companies (Table 5.10).

Additional information related to all these 27 students (who were at the end of their studies and ready to prepare their final thesis) participating in the internships are as follows:

- the total number of months for the internships (altogether): 34;

Table 5.10 Internships info

The sending university	No. of participants	Year/period	Internship place
Lebanese University, Beirut, Lebanon	3	2017	Romania
Notre Dame University, Lebanon	3	2017	Romania
University of Bahaman, Lebanon	2	2017	Romania
Beirut Arab University, Lebanon	2	2017	Romania
Lebanese University, Beirut, Lebanon	1	2017	France
University of Bahaman, Lebanon	1	2017	France
Lebanese University, Beirut, Lebanon	3	2018	Romania
Notre Dame University, Lebanon	2	2018	Romania
University of Balamand, Lebanon	4	2018	Romania
Beirut Arab University, Lebanon	4	2018	Romania
Lebanese University, Beirut, Lebanon	2	2018	Sweden

- in France and in Sweden, the students from internships participated in activities carried out in laboratories;
- in Romania the students from internships participated in activities carried out in the oil and gas companies, such as: OMV-Petrom SA, DOSCO Petroservice SRL, CONPET SA, ROMGAZ and Bonatti International.

In terms of internships quality and the satisfaction level, two categories of activities were carried out: checking the level of satisfaction of the participants in the internships and the analysis carried out independently by the profile companies regarding the way of involvement and acquiring knowledge from the participants in the internship. The verification of the satisfaction level of the participants in the internship was also done through a questionnaire designed by the KTH students, and the feedback was requested every time by the Romanian officials at the end of the internship.

At the same time, the companies in the Romanian oil and gas industry conducted their own surveys through which they wanted to obtain useful information for improving the activities related to the internship. It was desired to obtain a feedback showing the level of motivation of the participants, their level of involvement and the ability to acquire new specialized knowledge in a short period of time.

The questionnaire regarding the level of satisfaction of the participants in the internships aims also to obtain detailed information regarding the usefulness of the activities performed within the internship, the degree of closeness between the knowledge acquired in the university and the notions carried out in practice, the way of interacting with the designated persons from companies in the guidance of participants at the internship, the organization level for the internship, the updated content for the proposed internship agenda etc.

For this particular activity (i.e. internship) the last five questions from the questionnaire were selected (out of a total of 21 closed questions) due to their relevance in terms of satisfaction level. Furthermore, a useful information is to know that the

Table 5.11 Average scores for each relevant question versus internship (I)

Internship	I1	I2
Question	-2017-	-2018-
17	3.33	3.40
18	3.50	3.60
19	3.33	3.50
20	3.66	3.80
21	3.33	3.60

response rate for this questionnaire was 70.3% (19 students out of 27), which could be considered significant.

The percentage values were translated into scores associated with an adapted scale for quantifying the answers (using again a Likert scale with score 0 meaning totally unsatisfactory and score 4 which means totally satisfactory). This operation is presented in Table 5.11. These scores are good and very good. In addition, all the questions show an improvement in the grades given from I1 to I2. This can be explained by the comments received on the final open question. The participants underlined that the number of options regarding the company for the internship had increased (five companies in 2018 versus three companies in 2017).

Google Forms offers a lot of different ways to present processed data from questionnaires. Therefore, in this case, some processed data in the case of internships carried out in Romania in 2017 and 2018 will be presented. These data describe, using the *pie chart*, the overall percentage for the satisfaction level of the participants to these internships. The values are processed related only to the section that includes questions 17–21. These questions are considered, in this case, relevant because they refer to the supervision and guidance offered during the internship, to the formal assessment carried out for each participant at the end of the internship, to the possibility of obtaining new skills and knowledge during the internship, etc. So, the next two diagrams (*pie-chart* type) reflect this overall score for the internships (Figs. 5.3 and 5.4).

The feedback received each time was discussed at each PMB (Project Management Board) meeting (which took place approximately once every 5–6 months) and, practically, was continued through the exchange of emails.

Below on the next page is presented as an example, using Google Forms, the question number 20 and the answers to this question from the questionnaire addressed to the participants at the internships (Fig. 5.5).

Based on these discussions the board made proposals and recommendations for measures aimed at eliminating shortcomings, delays in implementing the project and correcting the results obtained during the course of the project. In this sense, the curriculum was corrected, the contents of the proposed disciplines were reconstructed, the topics discussed within the internship were reworked, the work plans for the internships were reconfigured and the plan for the in-situ places were rethought [39].

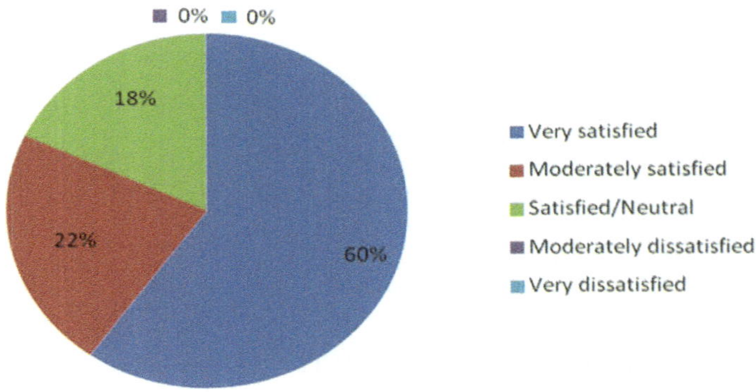

Fig. 5.3 Overall percentage for the satisfaction level in the case of internship 1

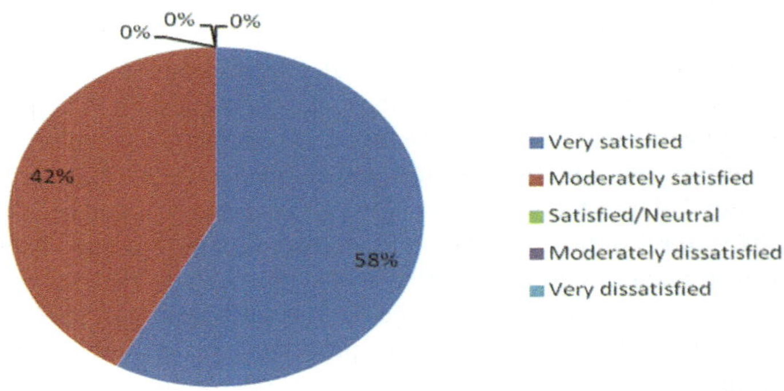

Fig. 5.4 Overall percentage for the satisfaction level in the case of internship 2

5.5 Discussion

The activities that were monitored, analyzed and assessed with an increased concern by the management project team were the workshops and internships (due to the fact that their implementation was plan to last a long time). However, no large differences in values were found between the same types of activities in the quantitative comparisons made and based on the processing of the questionnaires. In addition, the values obtained after processing these questionnaires are all placed above the minimum acceptable values initially set in the project. A first important idea resulting from these conclusions is that the organizational effort was very well carried out by the project management team. Moreover, assigned resources were well dimensioned in the project budget. These good results represented important premises for the set-up of the new Master's program starting with the academic year 2019–2020 at all four universities from Lebanon involved in the project.

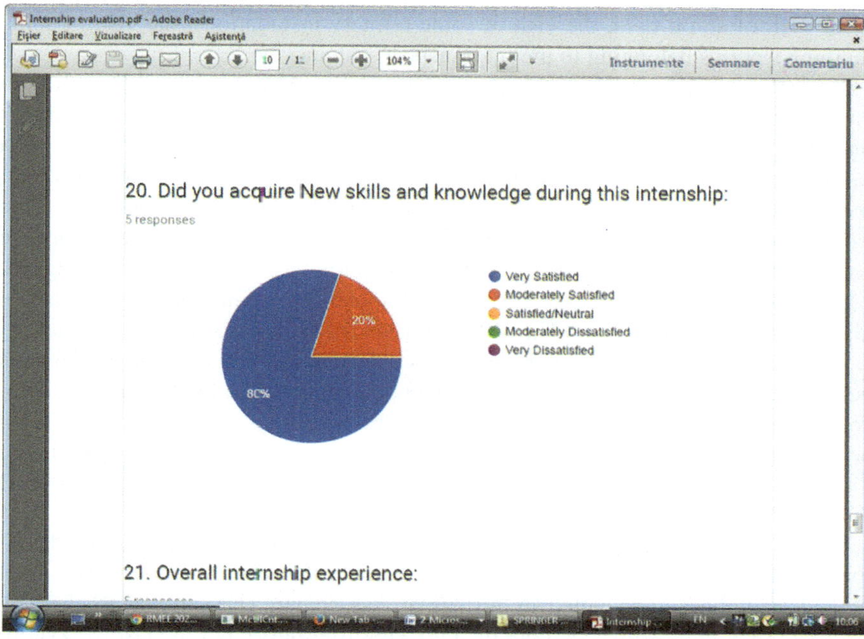

Fig. 5.5 Example of question and overall results (using *pie chart*) regarding the level of acquiring skills and knowledge (during the internship) in relation with participants' satisfaction—*capture from the main document*

The workshops were carried out primarily through Direct Instruction, by emphasizing broader concepts, as a whole, in order to ensure the understanding of these approaches. Then, depending on the situation, on the content of the discipline and on the trainer, other learning models were combined in order to verify the knowledge presented and discussed, using an efficient and objective assessment tool. Procedurally, each time at the end of the workshop period in the last hour participants were requested to complete online the questionnaire made in Google Forms. On the one hand, the Collaborative Projects model was used in the applications part (such as for the Petroleum Economic and Management or Reservoir simulation courses) by involving the participants in joint mini-project solving activities. On the other hand, the model of Practical learning activities was used, in which certain experiments were performed (as for example in the case of the Drilling & Production Engineering course, the subject Production and completion engineering or for the Refining processes courses). What should be noted is that the structure of the questionnaire allowed the mentioning of these learning methods used by one or another of the trainers when evaluating the workshop. Last but not least, in the agendas of these workshops in-situ visits were provided, and they are part of the Experiential learning model, as participants were able to see certain practical aspects, which were mentioned by trainers in the workshops through the Direct Instruction model (such as the visit of an oil field, the visit of some refinery installations, the visit of some pumping stations belonging to the oil transport systems, etc.).

By carrying out these questionnaires and processing the answers received, the aim was to verify the way in which the students or participants in the project activities managed to acquire most of the distributed knowledge. Each time, after carrying out a training or practice activity, discussions were organized by emails and PMBs (project management board) meetings, between the responsible persons belonging to all the project partners, through which proposals and action plans were made to improve these activities (agendas for workshops and discipline contents were redesigned, trainers' activity was discussed). The feedback sent by the trainees/students helped to adjust the training agendas, to identify better means of focusing on the trainee/student.

Thus, in the case of workshops, the things that were requested by the participants were in-situ visits in representative companies from both the upstream and downstream areas. In response to this complaint, the organizers of workshops 3 and 4 (which took place in Romania) modified the agenda for that week by introducing 2 additional visits to companies in the middle stream area (hydrocarbons transport) and in the downstream sector (i.e. crude oil processing).

A special remark concerns the content of the questionnaire regarding the development of the internships, which took into account the use of another assessment method, namely the one called Confirmative assessment. The aim was to check at the time of the internship (i.e. more than 90 days after the end of the last workshop) to what extent the issues addressed by the trainers are found in practice and at what level of compliance.

In the case of internships, starting from what was achieved in 2017, a second internship was organized in 2018, by eliminating some weaknesses (chosen topics, places to visit, must-see things), highlighted by the participants' questionnaires in 2017. On the other hand, in terms of internships, the main other grievances were related to the number of participants (a higher number of participants was requester), to the number of companies in the field that support internships (a higher number of companies was requested as well), to the desire to have internships organized in companies such as refineries.

The answers were processed quickly with the purpose to obtain useful information for carrying out all activities in the project in very good conditions. This mode of action, through questionnaires built through Google Forms, generated an efficient way of working between the project managers and its beneficiaries.

As can be seen from subsection 1.4.3, the number of people participating in the internship increased in the case of Romania from 10 (in 2017) to 13 (in 2018), and the number of companies that received internship students increased from 3 (in 2017) to 5 (in 2018). The only problem left unsolved was the impossibility of co-opting a refinery (as representing the downstream sector) in the partnership, so that at least 1–2 students could do internships there.

The questionnaires proved to be necessary and useful for the successful implementation of the described project. These questionnaires have proven their validity and reliability (in the case of the workshops the questionnaire was applied 5 times). They are credible as long as they were completed anonymously (thus, objectively), using an easy and fast communication environment (Google forms) and being processed by a small, but professional team. Questionnaires provided vital information to the project management team, presenting a high level of generality because these questionnaires are applicable to Master programs in various fields of study (especially engineering) and for various universities in Europe, Asia, etc.

One aspect that needs to be known concerns the validity and reliability of the questionnaires. The targeted questionnaire was the one used in the case of the workshops as it was used 5 times (during 5 workshops). Thus, for validity, a concurrent validity approach was taken into account in order to demonstrate the ability of the questionnaire to estimate the present performance of the beneficiaries. This proved the high level of information and knowledge reached by the participants through training during the workshops organized. Reliability verification focused on two forms: test–retest reliability, calculating the value of the Pearson correlation coefficient as 0.82 (this means good reliability); Inter-rater reliability, for which the value of the Cronbach correlation coefficient is 0.87, which demonstrates agreement between raters.

In order to clarify the content of this chapter and the fact that, in itself, the questionnaire did not only want to determine the level of satisfaction of beneficiaries of theoretical and practical training (this is in fact only a pillar of the questionnaire) it should be noted that the other pillars for the surveys refer to: organization of project activities, actual content/proposed agenda for each activity, methods and tools used in training, efficiency project implementation process.

The results, generated through the project implementation, largely confirmed the outcomes proposed by the project. On this occasion, a work procedure was verified and a workflow was built for the successful development of projects with similar content. Small corrections were made (mentioned in the chapter) despite a number of problems such as: a delay in funding, a longer discussion related to the completion of the curriculum, initial problems related to the organization procedures and the content of the presentations to be made in workshops; cultural differences; differences in the structure of the educational systems defining the participating countries; different systems on how to grant transferable credits for disciplines; the difficulty of concluding partnerships with companies that organized internships, differences in legislation regarding the accreditation process of academic study programs, etc.

5.6 Conclusions

This chapter aims to represent an element in the area of dissemination of project results and a guide of good practices in terms of project implementation. In the management of academic educational projects, a careful monitoring of the complex activities is necessary and it requires from the implementation project team a permanent effort to verify the compliance with the planned level of the results, in different phases of the project. This monitoring includes the interrogation of the project participants regarding the organization of the project activities, the continuous updating of the content of the activities for providing adequate training services (both theoretical and practical), the level of satisfaction of the beneficiaries etc.

Moreover, it was desired to show a useful way or tool to describe various means of obtaining a comprehensive and objective feedback regarding, for instance, the satisfaction of the different categories of beneficiaries integrated into the project.

The tool included in WP called *Quality Plan* related to the monitoring of the project implementation, the evaluation of the project progress obtained during the activities and the efficiency of the project activities was to use, on a large scale, the conduct of surveys based on online questionnaires.

From all available tools, Google Forms was chosen, taking into account the advantages described earlier in the chapter, to be used in the design and submission of all types of questionnaires. In order to have an objective and useful assessment, an e-assessment procedure was designed, organized, implemented and validated.

In this survey different means of presenting the results processed through Google Forms were used: spreadsheets, tables and pie-chart diagrams. In addition, they were used as elements to describe, to interpret adequately and easily, the feedback of the respondents in relation to the proposed assessment scales, percentages and scalar values/assessments.

Regarding the projected results, a minimum average score was set up for the satisfaction level of the participants in any type of planned activity (workshop, webinar, internship, etc.) of 3 out of maximum 4.

From a quantitative point of view in the case of feedback for the workshops there is a distribution of values recorded as follows: 1 value below level 3 (planned) i.e. 1%, 4 values of 3 (4%), 47 values between 3 and 3.49 (47% of appreciations), 18 values between 3.5 and 3.99 (18%) and 30 values of 4 (30%). Therefore the overall average score for the workshops is 3.55, while for the internships this score is 3.50. Both of these overall scores are much higher than the value initially set to be obtained in the sense of declaring the activities as beneficial to the project outcomes.

To conclude, this chapter is intended to be a guide of good practice in the successful implementation of projects aimed at introducing new study programs. The descriptions presented in different parts of the chapter explain in detail what needs to be done, what steps need to be taken, what are the procedures to follow, how to avoid certain implementation difficulties. All the project stakeholders involved had an important contribution to the success of the project so that anyone interested can learn about the design and operation of such an academic consortium and about how to implement a project with a similar goal. Last but not least, it is worth highlighting the key elements of the project, in fact the most successful aspects: the training of future trainers (let us not forget that this is a pioneering field in Lebanon), the establishment of partnerships with companies in the oil and gas industry useful for students and future graduates, but also the speed and efficiency in reconfiguring the agenda related to topics of theoretical interest to be discussed in the next workshops (as it was the case with W3 and W4, as compared to W2).

References

1. Cohen, L., Manion, L., & Morrison, K. (2007). *Research methods in education* (pp. 317–346). Oxford, UK: Routledge Publishers (See Questionnaires).
2. Gray, D. E. (2009). *Doing research in the real world* (pp. 187–211). London, SAGE Publications Ltd. (See Collecting Primary Data: Questionnaires).
3. Stone, D. H. (1993). Design a questionnaire. *British Medical Journal, 307*(6914), 1264–1266.
4. Lietz, P. (2010). Research into questionnaire design. *International Journal of Market Research, 52*(2), 249–272.
5. Sudman, S., & Blair, E. (1998). *Marketing research: A problem solving approach* (pp. 274–286). New York: McGraw Hill.
6. Lehmann, D. R., Gupta, S., & Steckel, J. H. (1998). *Marketing research* (pp. 183–190). New York: Addison–Wesley Educational Publishers Inc.
7. Saris, W. E., Gallhoher, I. N. (2014). *Design, evaluation, and analysis of questionnaires for survey research* (2nd Edn.). Hoboken: Wiley
8. Belson, W. A. (1981). *The design and understanding of survey questions.* Lexington Books
9. Converse, J. M., & Presser, S. (1986). *Survey questions: Handcrafting the standardized questionnaire* (no. 63). Sade Publications Ltd.
10. Rønningsbakk, L., Wu, T. T., Sandnes, F. E., Huang, Y. M. (Eds.) (2019). *Innovative technologies and learning: second international conference, ICITL 2019*, Tromso, Norway. December 2–5, Proceedings, Springer Nature Switzerland
11. Couper, M. P. (2000). Web surveys: A review of issues and approaches. *The Public Opinion Quarterly, 64*(4), 464–494.
12. Callegaro, M., Manfreda, K. L., & Vehovar, V. (2015). *Web survey methodology*. Sage Publishing Ltd.
13. Hippler, H. J., Schwarz, N., & Sudman, S. (Eds.) (2012). *Social information processing and survey methodology*. Springer Science & Business Media.
14. La Counte, S. (2019). *The ridiculously simple guide to google apps (G Suite): A practical guide to google drive google docs, google sheets, google slides, and google forms*. Independently Published
15. Ouyang, F., & Scharber, C. (2018). Adapting the TPACK Framework for online teaching within higher education. In P. A. Hershey (Ed.), *Teacher training and professional development: Concepts, methodologies, tools, and applications book* (pp. 1103–1120). IGI Global: Information Management Association.
16. Haddad, R. J., & Kalaani, Y. (2014). Google forms: A real-time formative assessment approach for adaptive learning. In *Proceedings of the 2014 American Society for Engineering Education, ASEE Annual Conference and Exposition Indianapolis, Indiana*; https://peer.asee.org/20540
17. Rebiere, O., & Rebiere, C. (2019) *Use Google forms for evaluation: Google forms and quizzes as effective educational tools, Guide Education Series*. Independently Published.
18. Hay, T. (2020). *The 2020 ultimate beginner's guide to google sheets: Practical guide to master the use of google sheet*. Independently Published.
19. Kiesler, S., & Sproull, L. S. (1986). Response effects in the electronic survey. *Public Opinion Quarterly, 50*, 402–413.
20. Brown, G., Bull, J., & Pendlebury, M. (1997). *Assessing student learning in higher education*. London, New York: Routhledge.
21. Gupta, A., Gupta, K., Joshi, A., & Sharma, D. (2019). Tools for e-assessment techniques in education: a review. In A. Azevedo & J. M. Azevedo (Eds.) *Handbook of research on e-assessment in higher education book* (pp. 28–52). IGI Global
22. González-Marcos, A., Alba-Elías, F., Ordieres-Meré, J. (2017). An online assessment and feedback approach in project management learning. In *Proceedings of the 9th International Conference on Computer Supported Education (CSEDU 2017)* (Vol. 2, pp. 69–77). doi: 10.5220/0006367500690077.

23. McLoughlin, C., & Luca, J. (2001). Quality in online delivery: what does it mean for assessment in e-learning environments? In *Meeting at the crossroads. Proceedings of the 18th Annual Conference of the Australasian Society for Computers in Learning in Tertiary Education. Melbourne, Australia*, 9–12 December 2001, https://ro.ecu.edu.au/ecuworks/4841
24. Reeves, T. C. (2000). Alternative assessment approaches for online learning environments in higher education. *Journal of Educational Computing Research, 23*(1), 101–111.
25. Sanjuan, A. G., & Froese, T. (2013). The application of project management standards and success factors to the development of a project management assessment tool. *Procedia—Social and Behavioral Sciences, 74,* 91–100.
26. Besner, C., & Hobbs, J. B. (2006). The perceived value and potential contribution of project management practices to project success. In *Paper presented at PMI®Research Conference: New Directions in Project Management, Montréal, Québec, Canada.* Newtown Square, PA: Project Management Institute.
27. Rashad, A. M., Youssif, A. A. A, Abdel-Ghafar, R. A., & Labib, A. E. (2008). E-assessment tool: A course assessment tool integrated into knowledge assessment. In M. Iskander (Eds.), *Innovative techniques in instruction technology, e-learning, e-assessment, and education.* Springer, Dordrecht.
28. Collis, B., & Moonen, J. (2001). *Flexible learning in digital world.* London: Kogan Page.
29. Wiggins, G. P. (1998). *Educative assessment.* San Francisco: Jossey Bass.
30. Kendle, A., & Northcote, M. (2000). The struggle for balance in the use of quantitative and qualitative online assessment tasks. In R. Sims, M. O'Reilly & S. Sawkins (Eds.), *Learning to choose, choosing to learn* (pp. 531–540). *Proceedings of the 17th annual Australian Society for Computers in Learning in Tertiary Education 2000 conference,* Southern Cross University, Coffs Harbour, Australia
31. Abbey, B. (2000). *Instructional and cognitive impacts of web-based education.* Hershey: Idea Group Publishing.
32. Greeno, J. P., & Hall, R. P. (1997). Practicing representation: Learning with and about presentational forms. *Phi Delta Kappan, 78*(5), 361–367.
33. Berge, Z. L., Collins, M., & Dougherty, K. (2000). Design guidelines for Web-based courses. In B. Abbey (Ed.), *Instructional and cognitive impacts of Web-based education* (pp. 32–41). Hershey: Idea group Publishing.
34. Lehnert, W. (1977). Human and computational question answering. *Cognitive Science, 1*(1), 47–73.
35. Wanous, J. P., & Lawler, E. E. (1972). Measurement and meaning of job satisfaction. *Journal of Applied Psychology, 56*(2), 95–105.
36. Saris, W. E., & Gallhoher, I. (2007). Estimation of the effects of measurement characteristics on the quality of survey questions. *Survey Research Methods, 1*(1), 29–43.
37. Smith, T. W. (1987). The art of asking questions, 1936–1985. *The Public Opinion Quarterly, 51,* 95–108.
38. Babo, R., Rodrigues, A. C., Lopes, C. T., de Oliveira, P. C., Queirós, R., & Pinto, M. (2012). Differences in internet and LMS usage a case study in higher education. In *Higher education institutions and learning management systems: Adoption and standardization* (pp. 247–270), IGI Global.
39. Mohanka, M. (2019). Guide to e-assessment with real time case studies & suggestive e-submissions. TAXMANN*'s Publishing.
40. Mohanka, M. (2019). https://www.google.com/intl/ro_ro/forms/about, Retrieved on April 9, 2019.
41. Mohanka, M. (2019). https://www.fractuslearning.com/using-google-forms-assessment, Retrieved on September 20, 2019.

Part II
E-Assessment Approaches

Chapter 6
FLEX: A BYOD Approach to Electronic Examinations

Bastian Küppers and Ulrik Schroeder

Abstract Electronic assessments are a topic of growing importance for universities. Even though it is not a new approach, there still is no "common way" for implementing these assessments. Often times, the assessment is conducted in a computer lab of the educating institution. Still, this approach has some drawbacks, for example the costs and management overhead introduced with operating a computer lab. However, the biggest drawback arises for the students, as they have to use unfamiliar devices. If a group of students has to work with an unfamiliar (software) environment in the assessment, for example a Windows system when using MacOS otherwise, this limits these students in their effectiveness during the assessment. Students that are used to work with a Windows system do not have this handicap utilizing a BYOD approach, i.e. allowing the students to use their own devices in an electronic assessment, is a solution to the mentioned problem. However, a BYOD approach does not only solve problems, but at the same time introduces new challenges that have to be tackled. Most notably, cheating prevention is a topic of high relevance. Additionally, differences between the students' devices must not lead to unequal opportunities for the students in the assessment. This chapter describes the implementation of a software framework for BYOD electronic assessments and discusses novel solutions to the issues related to BYOD. Reliability and security are in this context topics of special interest, as students' devices have to be considered as untrusted devices in general. Therefore, an integral part of the solution is a way to detect modification to the examination client.

Keywords Electronic examinations · Computer based assessment · Computer aided assessment · Technology enhanced assessment · Bring your own device

---

B. Küppers (✉) · U. Schroeder
Learning Technologies Research Group, RWTH Aachen University, Ahornstraße 55, Aachen 52074, Germany
e-mail: bastian.kueppers@rwth-aachen.de

U. Schroeder
e-mail: schroeder@informatik.rwth-aachen.de

R. Babo et al. (eds.), *Workgroups eAssessment: Planning, Implementing and Analysing Frameworks*, Intelligent Systems Reference Library 199,
https://doi.org/10.1007/978-981-15-9908-8_6

6.1 Introduction

Following the general trend of digitization, more and more digital elements are being used in university teaching at German universities, for example learning management systems (LMS) or mobile apps [1]. The integration of digital elements covers the entire spectrum of scientific events, with one exception: examinations [2]. Similar trends can also be observed in other countries like the United Kingdom [3], Greece [4], the United States of America [5] or Australia [6]. Due to the slow progress in integrating digital exams into higher education, there is a significant media gap between lectures and exercises containing digital elements and exams that are still mainly written in analogue form, i.e. on paper. The retention of paper examinations is often caused by reservations about digital examinations. Such reservations include the fairness of digital examinations or the reliability of digital examination systems [7]. However, digital examinations have significant advantages that make them worth considering. Sometimes the reservations about e-Assessment are also based on prejudices that may or may not be justified. There are also financial reasons for not introducing digital examinations, as it is costly to maintain a centrally managed IT infrastructure for digital examinations. The latter is reported, for example, by several German universities that have such an infrastructure, such as the University of Duisburg-Essen [8] and the University of Bremen [9]. Since most students already have equipment suitable for digital examinations [10–12], Bring Your Own Device (BYOD), an emerging trend in higher education [13, 14], is a possible solution to overcome these problems. Beyond this rationale, BYOD ensures a familiar working environment for students, which is another advantage for electronic examinations (EA) in a BYOD scenario compared to a centrally managed IT infrastructure. From the students' point of view, this is the biggest advantage, as an unfamiliar working environment can be a handicap for them. For example, if a student is used to a certain operating system (OS), a device with another OS can be an obstacle because it works differently, e.g. keyboard shortcuts have an unexpected behavior. A more detailed discussion of the prejudices and advantages of EA can be found in [15].

With this in mind, our goal was to develop a framework for EA with a BYOD approach. Framework in this context means a software for conducting the exam as well as a set of formal requirements that a higher education institute (HEI) must meet to ensure the reliability of the exam beyond the sole reliability of the exam software.

Before we could start development, however, there were still some open questions that had to be clarified. In particular, *Safety* and *Quality of Treatment* are issues that must be considered when dealing with EA. In addition, it is very important to consider the requirements of all parties involved. These are the examiners, the students and the administration at an HEI. All these parties have requirements, either by personal preference or by laws or other regulations, which must be observed for a successful solution.

Throughout this chapter, some terms will appear quite frequently. For the ease of readability, throughout this chapter the following wording will be used.

EA framework	The unity of all pieces of software and formal requirements.
EA software	The unity of all pieces of software (server and client).
EA application	The software that runs on the students' devices (client).
EA server	The complete server infrastructure to which the EA application connects.

As the research project described in this chapter was carried out during a PhD project, parts of this chapter have been taken from the PhD thesis of Bastian Küppers both, figuratively and literally.

6.1.1 Research Questions

The development of a BYOD solution for EA was guided by the following research questions. They mainly concern the previously mentioned fields *Security* and *Equality of Treatment*.

1. How do personal attributes influence the perception of e-Assessment?
2. How can e-Assessment on students' devices be reliably secured in a BYOD setting?
3. How can the authorship of examination results be attributed to a particular student?
4. How can the integrity of examination results be ensured?
5. How can students be treated fairly and equally in a BYOD setting?

6.1.2 Research Methodology

Guided by the research questions, the actual research was conducted in a *Design-based Research* process as defined by Wang and Hannafin in their work *Using Design-based Research in Design and Research of Technology-Enhanced Learning Environments* [16] as follows

> [Design-based research is a research] methodology aimed to improve educational practices through systematic, flexible, and iterative review, analysis, design, development, and implementation, based upon collaboration among researchers and practitioners in real-world settings, and leading to design principles or theories.
> Wang and Hannafin [16, pp. 6–7]

Beyond this definition, March and Smith came to the conclusion that "*[N]atural science tries to understand reality, [whereas] design science attempts to create things that serve human purposes*" [17, p. 253]. Which brought Hevner et. al to the insight that "*Design science [...] creates and evaluates IT artifacts intended to solve identified organizational problems*" [18, p. 77]. Therefore this research paradigm was chosen because an EA framework is not something natural that can be discovered, explained and understood, but something that needs to be developed to serve a specific purpose.

6.2 State of the Art

Having a closer look at potential ways of implementing BYOD approaches for EA, it turns out the differences between these approaches mostly concern two major aspects:

- Which software is used on the students' devices?
- How are the students' devices connected?

Basically, the students' devices can either be used as a workstation or as a (thin) client for a hosted infrastructure [19]. In addition, the students devices' can either be connected via ethernet, via Wi-Fi or not at all.

When using the students' devices as workstations, the software used during the EA is installed and executed on the student devices. There are different approaches to this, which differ in the degree of freedom of the students and in the possibilities to take precautions against fraud [20]. In order to make things more manageable for the examiner, students can be required to use specific software provided by the examiner. However, obliging students to use specific software during an EA, e.g. NetBeans for a programming course, does not provide any security. Therefore, along with the required tools, a so-called *LockDown* software is often required by the examiner, e.g. *Safe Exam Browser*.[1] However, this approach raises potential problems regarding software compatibility, as all students must be able to run the *LockDown* software on their devices. This can become a problem if a student's operating system is not supported by the software. Another approach to implementing a BYOD approach to EA is to use the students' devices as Thin Clients. This means that these devices are not used to work on them, but only as a client to connect to a central IT infrastructure, such as a remote desktop server. In such scenarios, the examiners have full control over the work environment itself and can provide a preconfigured environment. In addition, the student's results are available to the examiner on the remote server and do not have to be retrieved via USB sticks or similar. However, this scenario is susceptible to fraud, as students may be able to work outside their thin clients if no precautions are taken. However, these precautions will cause the same problems as those already discussed for using the students' devices as workstations.

With regard to the connection of the students' devices, there are technically more possibilities than mentioned above, e.g. a connection via a cell network. However, this type of connection is not preferable for EA, as it is not under the examiner's control. Depending on the BYOD approach chosen, it is not necessary to connect

[1] https://safeexambrowser.org/.

the students' devices, e.g. if the students can choose the software to work with themselves or if a complete operating system has been provided on a USB stick, which is also used to collect the results. However, the connection of student devices may be necessary if the students' results are to be delivered via a LMS or if online materials have to be accessible. If the student devices are connected to a local network, either Ethernet or Wi-Fi, this network can allow Internet access controlled by a firewall. Whether a connection to the Internet is necessary depends on the chosen approach to EA. If student devices are used as thin clients, there is no need to connect to the Internet, but a connection to the local network is sufficient. However, connecting to the local network does not prevent students from communicating per se. To prevent students from doing so, the network must be configured accordingly. In addition, other communication methods, such as Bluetooth or LTE, have to be disabled The latter can be prevented by building structural measures, for example by constructing examination rooms that resemble a Faraday cage. Locally established Bluetooth connections could be detected by the testers by searching for such connections themselves.

More details on the different approaches of realizing a BYOD approach to EA are available in [21, 22].

6.3 Requirements Engineering

The implementation of EA, as already discussed in Sect. 6.1, is often hindered by prejudices against EA both by examiners and students. However, it is essential that both sides accept EA as part of the HEI examination system [4]. It is therefore important to gather and analyze the concerns of examiners and students in order to develop EA software that is widely accepted and successfully used at the HEI. This process is very similar to *Requirements Engineering* in software engineering. This is not surprising, since software development is an essential part of the EA framework development. According to Lukarov, "*three main groups of stakeholders and decision-makers [exist]: students and faculty, and administrative bodies*" [23]. The requirements imposed by these parties are discussed in the next sections. At this point it is important to mention that requirements engineering was carried out within the context of the MATSE educational program at RWTH Aachen University. MATSE is the abbreviation for ***MA****thematical and* ***T****echnical* ***S****oftware* ***DE****veloper*. This training programme includes vocational training in a research institution or company and the bachelor's degree programme *Angewandte Mathematik und Informatik* (Applied Mathematics and Computer Science, formerly known as *Science Programming*) at FH Aachen University of Applied Sciences [24]. Nevertheless, the gathered requirements in this special environment were identified as generally valid. However, the EA framework was developed with the thought of flexibility in case of missing requirements.

6.3.1 Administrative Bodies

At RWTH Aachen University, administrative bodies have publicly announced which requirements they deem important for EA as part of the "*Digitalisierungsstrategie der Lehre*" (digitalization strategy for teaching) [25, 26]. Additionally, the data privacy requirements at RWTH Aachen University were published in the "*E-Learning Ordnung*" (e-learning ordinance) [27].

In the first two documents, boundary conditions for EA at RWTH Aachen University are formulated. It is stated that "a concept for providing the necessary infrastructure for the required blended learning activities is agreed upon and permanent financing is guaranteed. The infrastructure includes [...] Wi-Fi [and] the possibilities to carry out electronic examinations including the equipment and necessary hardware" [26, p. 7].

The e-learning ordinance states that "responsible persons are allowed to process personal data of the users when utilizing e-learning systems, if this ordinance or other legal regulations explicitly allow it" [27, p. 3]. However, "the use of e-learning systems has to be aligned to the goal of collecting and processing as little personal data as possible" [27, p. 3]. This means that the students have to be able to use an e-learning system anonymously, if that does not contradict the purpose of the system [27, p. 4]r. Persons who are responsible for an e-learning system have to provide the students with a short description of the collection and usage of personal data [27, p. 4]. The students have to explicitly accept these conditions, which possibly has to be done electronically, and the acceptance has to be stored reliably [27, p. 5]. Additionally, the e-learning ordinance states explicitly which data security measures have to be taken for an e-learning system. The first subsection defines that "responsible persons have to take technical and organizational measures to prevent the abuse of the personal data that was collected and used on the basis of this ordinance" [27, p. 6]. Therefore, appropriate measures have to be taken to ensure that ...

1. ..."the purpose limitation of the collected data is warranted" [27, p. 6]
2. ..."only authorized persons can access the data for which they have authorized access and personal data can not be read, copied, modified or deleted without authorization" [27, p. 6]
3. ..."it can be subsequently verified and established whether and by whom personal data have been entered, modified or erased in data processing systems and to whom they have been disclosed" [27, p. 6]
4. ..."personal data is protected against accidental deletion or loss" [27, p. 6].

6.3.2 Students

Existing literature suggests that students generally have a positive attitude regarding EA [28–44]. However, most of these studies focus on a specific HEI or even on a single course of study. Furthermore, most of the available literature does not deal

Table 6.1 Demographics of the participating students

	Male (%)	Female (%)	[NA] (%)	Σ (%)
<18 Years	1.23	0.25	0	1.48
18–25 Years	60.29	16.66	0.49	77.44
>25 Years	13.97	6.37	0	20.34
[NA]	0.49	0	0.25	0.74
Σ	75.98	23.28	0.74	100

with EA in a BYOD setting. Therefore, we have conducted a survey to determine the specific requirements for this approach. To get a broader picture than the existing surveys can deliver, we invited students from several HEIs to participate in the survey, namely from RWTH Aachen University, FH Aachen, Alpen-Adria-University Klagenfurt, TU Berlin, FOM Hochschule für Oekonomie und Management (Study Center Aachen) and University Albstadt-Sigmaringen. A total of 408 students took part in the survey, which was carried out as an online survey using the tool *SoSciSurvey*.[2] Therefore, every student could take part in the survey, without the need to be present in a lecture hall at a given time or the like. The link was provided to the students via the mailing lists at the different HEIs. The demographics of the participating student are shown in Table 6.1.

About three quarters of the participating students are male and a little more than one fifth is female. A similar distribution can be observed for the age, where about three quarters are aged between 18 and 25 years and a fifth of the students is aged above 25 years. The students study in a variety of study programs, including Applied Mathematics and Computer Science, Artificial Intelligence, Data Science for Decision Making, Engineering and Physics. Despite some students studying in programs like economics or literature, the vast majority of the study programs is related to a *STEM* topic. Additionally, a high technology affinity throughout the participants of the survey is indicated. Since too few participating students were enrolled in a study course that is not from the *STEM* field, the collected data is not suitable to answer whether the study course influences the students' perception of EA. It may be speculated that the absence of those students may be caused by the decision to carry out the survey via an online portal, which may have biased the results in a way that only students participated who have an affinity for technology [45]. However, that can not be concluded from the data, as results from a comparable survey with additional modes of participation are not available.

To examine the influence of *gender*, *age*, *technology affinity* and *study level*, the data set was accordingly split into subsets which where then compared to each other for statistically significant differences. The comparison was carried out for the six questions **E1**, **E2**, **E3**, **B1**, **C1**, and **C2** of the students' questionnaire, which deal with the view on *e-Assessment*, *BYOD* and *Cheating*. This means that the data was split according to the options available for these particular questions. However, to

[2] https://www.soscisurvey.de/.

Table 6.2 *p* for the χ^2 Test for the Students

	Gender	Age	Technology affinity	Study level
E1	*0.039*$< \alpha_{0.05}$	0.11	*0.0004* $< \alpha_{0.01}$	0.18
E2	0.41	*0.002* $< \alpha_{0.01}$	*0.042* $< \alpha_{0.05}$	0.07
E3	0.60	0.14	*0.013* $< \alpha_{0.05}$	0.72
B1	0.20	0.20	*0.004* $< \alpha_{0.01}$	0.19
C1	0.55	0.62	0.63	0.32
C2	0.54	0.52	0.18	0.19

The italics indicate a value in the table which was a statistically significant value

determine the affinity for technology, the *TA-EG* questionnaire [46] has been used, which consists of 19 items to take into account, not just a single item. Therefore, a cluster analysis using a k-medoids algorithm [47] was performed to be able to construct valid subsets for the affinity of technology. These subsets were then analyzed for significant differences with a χ^2 *Test for* $r \times c$ *Tables* [48, pp. 493–572]. The resulting *p*-values for the Likert-scaled items can be found in Table 6.2.

Given these *p*-values, it is possible to take conclusions on the influence of *gender*, *age*, *technology affinity* and *study level* to a certain extent. Regarding question **E1**, it seems that women are more hesitant to accept EA as part of an HEI's examination system. In addition, students between the age of 18 and 25 years seem to be more positive about EA than students of other ages (**E2**). The cluster analysis revealed two clusters for the affinity of technology. These clusters represent one group that has clearly a high affinity to technology (*Cluster 1*) and another, more reserved group (*Cluster 0*). For these clusters, statistically significant differences between the clusters were found for questions **E1**, **E2**, **E3** and **B1**. The participants in *Cluster 1* have a more positive attitude towards EA and BYOD compared to the participants in *Cluster 0* The achieved results from the survey show a rather clear picture: The students would have preferred to have electronic examinations in their study programs. However, survey participants do not want it necessarily as a replacement for PBE, but rather as a complementary approach, as the answers for questions **E2** and **E3** suggest. This perception of EA seems to be governed by the advantages it offers, covering topics like *fast correction*, *more realistic assignments*, *more diverse examination tasks*, and *readability*. The latter was stated frequently in the free text comments during the survey. However, students are also concerned about disadvantages, like *security*, *usability*, and *fairness*. Additionally, technical difficulties and the subsequent loss of already solved assignments are very often mentioned in the comments. Overall, less than half of the students mention disadvantages in EA. However, especially when it comes to a BYOD approach, the students are afraid that technical difficulties may lead to a handicap for them or that they have to have a capable device on their own. Still, there is a positive tendency regarding a BYOD approach, as students see the advantage of a *familiar device in the examination*. Additionally, the topic of fairness is important to the students, as the state *differences between the students' devices* as the main concern when utilizing BYOD. Furthermore, topics like *security* and *cheating* are of importance for the students. The students are rather split about the risk of cheating in PBE, yet there is a tendency that students think that it is easier to cheat in EA. The age seemingly being a factor that influences the perception of EA is in line with the concept of *Digital Natives* introduced by Prensky [49]. He claims that "*[t]oday's students have not just changed incrementally from those of the past*", but underwent a drastic change of attitude, because "*the arrival and rapid dissemination of digital technology in the last decades of the 20th century [was] an event which changes things so fundamentally that there is absolutely no going back*". The evidence gained from the survey suggests a similar conclusion, because there is a statistically significant difference between students over the age of 25 years in comparison to younger students. This is in line with the definition of *Generation Z*, which is according to Anthony Turner "*born from 1993 to 2005*" [50]. The timespan mentioned in Turner's article is exactly in line with the findings of the age that does have an influence on the perception of EA. Gender having an influence on the perception of EA is actually not surprising, as many studies show that women seem to have a lower confidence in using technology in general than men, for example by Kadijevich [51], Kahveci [52], and Yau and Cheng [53], whether this is justified or not. Therefore, it is reasonable to assume that the same tendency can be observed when examining the perception of EA. It is not surprising either that technology affinity has influence on the students' perception of EA, since EA uses technology to carry out a previously analog process. Similar trends have already been observed in other areas, for example in medicine, where telemedicine was the new, digital variant of a previously analog process, [54] or museum tours, where the human guide was replaced by a digital device [55]. Therefore, it can be concluded that students who show affinity to technology are more positive towards EA than students who have no favorable, or even an unfavorable, attitude towards technology.

Additionally, the data collected points to several aspects that the students perceive as critical. The students consider many characteristics of EA and BYOD as quite positive. However, the survey participants also express a variety of concerns such as system crashes and the associated loss of data or general concerns about security and fraud. This is consistent with other studies by Kocdar et al. [56] and Hillier et al. [57]. Based on all of these aspects, which are perceived by students as positive or negative, it is important to address both sides in the concept of a framework to ensure that students accept EA in a BYOD setting. However, some aspects cannot be influenced by a framework, e.g. the perception of the mode of writing in this examination form. The data collected leads to the following list of important aspects to be considered:

1. Plans have to be available to deal with …
    a. …students who do not own a suitable device.
    b. …technical problems throughout the exam.

2. The EA software has to …
   a. …offer a good usability.
   b. …be freely available so that students can get used to it prior to the exam.
   c. …verifiable by the students.
3. Measures have to be taken to prevent …
   a. …unfair advantages for particular students.
   b. …cheating during the EA.
   c. …manipulation of the students' answers.
   d. …data loss during the exam, e.g. due to a system crash.
4. The whole EA process has to be transparent to the students to clear up doubts.

More details on the survey amongst the students can be found in [58]. The survey itself can be found in the appendix of this chapter.

6.3.3 Examiners

According to the available literature the attitude of the examiners towards EA is rather reserved [59]. A main reason for this is the possibilities of fraud, which will probably increase for EA [60]. In order to get a clearer picture of the origins of the examiners' cautious attitude, a survey was conducted to understand the examiners' views on EA and BYOD and to understand what factors influence them. A total of 110 teachers at RWTH Aachen University and FH Aachen University of Applied Sciences took part in the survey, which was carried out as an online survey using the tool *SoSciSurvey*.[3] The link to the survey was provided by the vice-rector for teaching at RWTH Aachen University and by the dean's assistant of the The Faculty of Medical Engineering and Technomathematics at FH Aachen University of Applied Sciences. The demographics of the participating teachers are shown in Table 6.3.

About three quarters of the participating teachers are male, a little less than one fifth is female and about five percent consider themselves none of the former. Regarding the age distribution, about five percent are younger than 30, two fifths are between 30 and 50 years old and the rest is older than 50 years. The teaching areas of the participants span a variety of fields. However, the majority (**68.17 %**) of the participants teaches a *STEM* topic and has been teaching for over 20 semesters. As with the survey that was carried out amongst students, the affinity to technology of the examiners was measured by the *TA-EG* questionnaire [46]. Again, the questionnaire covers a suitable set of features for performing a cluster analysis on. As with the results of the student survey, the data set was split into subsets to determine the influence *gender*, *age*, *field of expertise*, *teaching experience* and *institution*. In the same manner as for the students, these subsets were then analyzed for significant

[3]https://www.soscisurvey.de/.

Table 6.3 Demographics of the participating examiners

	Male (%)	Female (%)	Diverse (%)	Σ (%)
<30 Years	0.9	2.73	0	3.63
30–50 Years	26.36	11.82	1.82	40
>50 Years	50	4.55	1.82	56.37
Σ	77.26	19.1	3.64	100

Table 6.4 p for the χ^2 Test for the examiners

	1–5 S. (%)	6–10 S. (%)	11–20 S. (%)	>20 S. (%)	[NA] (%)	Σ (%)
Engineering	4.55	7.23	4.55	19.1	0.91	36.34
Medicine	0.91	0	0.91	10	0	11.82
Sciences	0	0.91	7.27	23.65	0	31.83
Social sciences	0	0.91	4.55	4.55	0	10.01
Other	0	0	0	2.73	0	2.73
Humanities	0	0	0	6.36	0	6.36
[NA]	0	0	0.91	0	0	0.91
Σ	5.46	9.05	18.19	66.39	0.91	100

differences with a χ^2 *Test for $r \times c$ Tables* [48, pp. 493–572]. The resulting p-values for the Likert-scaled items can be found in Table 6.4.

Given these p-values, conclusions on the influence of *gender*, *age*, *technology affinity*, *field of expertise*, *teaching experience* and *institution* can be drawn to a certain extent. In general, the tendency is the same as for the students: *gender* and *age* have an influence, however, on other parts of the survey. For the students, these demographic characteristics affected the answers to questions **E1** and **E2**, for the examiners the affected questions are **E3** and **B1**. The data indicate that male teachers are more convinced that a BYOD approach can be beneficial for assessment than other groups. The age of the examiners influences their view on question **E3**. The data suggests that with increasing age, the preconceptions regarding EA rise as well. For question **E3**, the distribution of the age leads to the conclusion that examiners of ages above 50 do not see EA as a suitable replacement for PBE, whereas younger teachers have a rather positive view on EA replacing PBE. As for the students, the cluster analysis revealed two clusters that were derived using a *k-medoids* clustering algorithm [47]. These clusters represent one group that clearly has high affinity to technology (*Cluster 0*) and another, more reserved group (*Cluster 1*). For these clusters, statistically significant differences between the clusters were found for question **E1**. The participants in *Cluster 0* have a more positive attitude towards EA than the participants in *Cluster 1*. The *field of expertise* has a significant influence on question **E3**. Teachers from the fields of *medicine* and *social Sciences* see EA as a suitable replacement for PBE. Teachers from other fields (*humanities*, *engineering*, *sciences*) are more reluctant regarding EA replacing PBE. The reason for this might well be the present examination policies in these fields. For example, as "*[m]ultiple-choice questions [...] are still used in high stakes exams worldwide to assess the knowledge of medical students*" [61, p. 1], it is easy to see how teachers in medicine using this mode of examinations perceive it as suitably replaceable by EA. The examiners generally agree that EA is a good supplement to PBE, while their estimate on the opportunities of cheating does not show a clear tendency. However, there seems to be a consensus that it is easier to cheat in EA than in a PBE. This in line with finding from the literature [62, 63].

The data collected leads to several aspects that the auditors consider to be critical. The examiners see several positive aspects of EA, such as better *exam management*, including student administration and statistics, and *innovative exercises*, possibly including multimedia. On the other hand, examiners are concerned about certain exercises or even entire exams that are not suitable for an electronic environment, or problems with organizational matters. Overall, the positive and negative aspects of EA balance each other out, but when it comes to EA in a BYOD setting, the examiners clearly reject this idea. Almost no positive aspects were mentioned by the examiners, and most negative aspects concern the possibility of fraud. It is therefore very important to address these aspects in the concept of a framework to convince the examiners that EA can be successful in a BYOD setting without opening up to fraud. From the data collected, the following list of important aspects to be considered has been derived:

1. The EA software has to ...
    a. ...offer a way to carry out management comfortably (register students, create assignments, correct, archive, ...).
    b. ...offer a software interface to implement and adapt types of assignments to the examiners' needs.
2. Measures have to be taken to prevent ...
    a. ...unfair advantages for particular students.
    b. ...cheating during the EA.
    c. ...manipulation of the students' answers.
    d. ...data loss during the exam, e.g. due to a system crash.
3. The whole EA process has to be transparent to the examiners to clear up doubts.

More details on the survey amongst the examiners can be found in [64]. The survey itself can be found in the appendix of this chapter.

6.3.4 Threat Model and Security Requirements

Security threats exist for every form of examination and they are exploited by the students [65]. In the implementation of EA the use of computers opens up new risks of fraud—in theory. This potential threat seems to worsen in a BYOD setting, as students' devices can be considered as *untrusted devices* in general. Beyond this reasoning, however, it seems reasonable to assume that fraud is something that must have a good cost-benefit ratio. Therefore, students would choose to cheat in a way that gives them the best possible advantage during the exam, provided that the type of cheat chosen does not lead to too much effort and, perhaps more importantly, to a higher risk of being caught. In some ways, it is still easier to use a smartphone in the toilet than to reverse engineer and hack digital examination software. However, the fraud hurdle should have very similar characteristics to the fraud itself: It should have good value for effort. Since absolute security is not possible, as pointed out by the security expert Schneier [66, 67], this is not the target to aim for. On the other hand, it seems unacceptable for examiners and administrators not to implement security measures. Therefore, the right level of measures must be defined which, on the one hand, does not require too much effort for development and operation, but, on the other hand, provides adequate security. This in turn is based on the assumption that EA must provide a similar level of security as paper-based examinations. However, in order to be able to derive meaningful security requirements, a threat model has first to be developed, which can serve as a basis for reasoning.

Based on work by Sindre and Vegendla [68], a threat model was developed which consists of the following threats:

1. Impersonation
2. Assistance/Collaboration
3. Plagiarism
4. Using Aids Not Allowed for the Exam
5. Timing Violations
6. Lying to Proctors
7. Smuggling Out the Exam Questions After the Exam
8. Manipulation of Exam Results.

Based on these threats, security requirements could be derived that a software solution needs to fulfill in order to be robust against the identified threats. For each of the identified threats counter measures could be identified. At this point it is important to note that these counter measures are not exclusively of technical nature, but in some cases are formal requirements that have to be implemented into the HEI's processes regarding examinations. In particular, the following requirements were derived:

1. The EA software has to …
    a. …automatically label students' results.
    b. …be able to authenticate students.

   c. …sign the students' results with a digital certificate.
   d. …provide the examination's exercises only in a limited timeframe.
   e. …collect the students' results at the end of the examination.
   f. …prevent the students from copying the exam questions.
   g. …give out receipts to the students to prove that they handed in their results.

2. Measures have to be taken to prevent …

   a. …plagiarism.
   b. …manipulation of the exam software.
   c. …prohibited actions during the examination.

6.3.5 Technical Requirements

Some of the collected requirements have a significant influence on the software architecture as well as on the choice of programming language.

To balance the differences between the devices, a concept called *computational offloading*, which originates from the field of mobile computing, can be used [69, 70]. This approach uses a remote system that provides enough processing capabilities to perform computationally intensive tasks. Although designed for smartphones and tablets, this principle can also be applied to notebooks and other mobile devices. For EA, computational offloading effectively reduces disparities between students' devices because all students rely on the same remote computer for computationally intensive tasks. The architecture of the EA software must be designed to allow for computational offloading. Similarly, the architecture must be able to perform remote maintenance for the EA client on the students' machines.

There are several ways to support multiple operating systems with one software. One possibility would be to create different source codes for the different operating systems. On the pro side, this allows to fine tune the different programs to the target operating system. The different source codes could even be written in different programming languages. This approach thus allows maximum flexibility in the development process. However, this approach is associated with a high level of additional effort, since each new function must be implemented separately for different operating systems. This is possibly the reason why the existing *LockDown* software is not available for every operating system. As this approach is too much effort, it is not economical and a different solution is needed, for example a programming framework that is available for the intended operating systems.

6.4 Implementation

Based on the requirements gathered during the requirements engineering the **FLEX** system was developed (Framework **F**or **FLE**xible **E**lectronic **EX**aminations). The basic architecture of the EA framework is depicted in Fig. 6.1. The four components of this architecture will be discussed in the next sections.

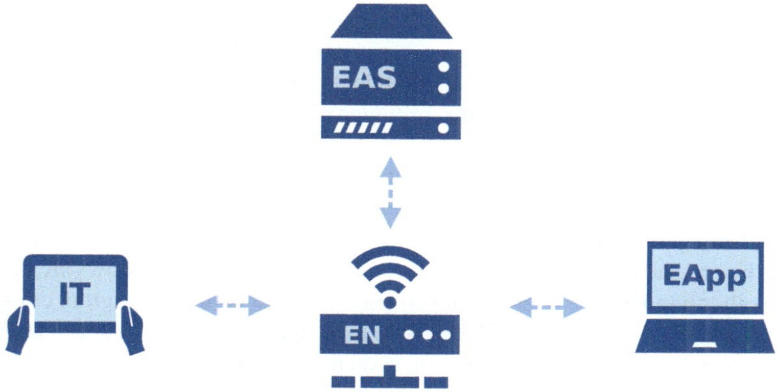

Fig. 6.1 Basic architecture of the EA framework

6.4.1 Technical Solutions

6.4.1.1 Prerequisites

To be able to counter the threats identified in the threat model, two techniques turned out to be crucial: *Remote Attestation* (RA) and *Code Obfuscation* (CO).

RA is a concept that can remotely determine the integrity of a piece of software which is executed on an untrusted device. Basically, it implements a challenge-response-protocol between the challenged system and a trusted server component using small executables called *agents*. Most of the available approaches are hardware-based, i.e. rely on specifics of the available hardware components. However, in a BYOD setting no assumptions regarding the hardware available in the students' devices can be made. Therefore, software-based approaches to RA have been taken into account, with Pioneer [71] being one of the first approaches. TEAS [72] improves the challenge-response protocol of Pioneer by using multiple verification challenges and introducing code obfuscation to these in order to prevent reverse engineering.

CO is a technique that disguises the inner workings of a program by modifying the program code in such a way that the functionality of the program is preserved, but it is almost impossible for a human to read it. Even if CO cannot prevent a program from being reverse-engineered in the long run [73], CO can considerably slow down the process of reverse engineering for a computationally limited opponent, such as a student during an exam. Therefore, CO in combination with RA is a valuable technique, especially in the context of EA, which offers an opponent only a narrow time frame for a successful attack. A comprehensive overview of concealment techniques and related mechanisms is available in [74, 75].

6.4.1.2 EA Application

The EA application is executed on the students' devices. It provides the interface for solving the tasks of the EA. It also provides a recognition mechanism to identify cheating-related actions on the devices.

The EA application is started in full screen mode, i.e. not overlaid by another window. It monitors from its start whether the student leaves the full screen mode and whether a part or the whole window is overlapped by another window. This could be done by processing the native *paint* event. If no overlapping windows are detected then no other window has been displayed and therefore no other program consisting of a graphical user interface (GUI) has been used apart from the EA application. If a student has a device with more than one screen, all additional screens will be filled with an empty window, allowing the same monitoring to take place. In addition, keystrokes are checked to ensure that they are really coming from a keyboard and not from another process that simulates keystrokes or from content pasted from the clipboard. Since there are valid use cases for the clipboard, a separate clipboard implementation must be provided, which is fully monitored by the EA application. The mouse is also monitored to ensure that the mouse pointer is not used by a background process to establish any sort of communication.

However, the approach described above only works if the message handling of the operating system is not manipulated, because in this case a student could prevent important events from being triggered. It must therefore be ensured that the processing of the necessary events works as intended. This is done by a runtime verification included in the agent of the RA.

The front end of the EA application is implemented with JavaScript (JS) and the *Electron* framework,[4] which is based on NodeJS.[5] The Electron Framework provides the ability to develop cross-platform applications that support the major operating systems and use web technologies in a client-server architecture. However, it is also possible to use native functions of the operating system, which is important for the integration of security functions into the EA application. If the provided functionality of the EA application does not support certain operations, the runtime environment can be extended by plug-ins provided in the form of shared libraries. These shared libraries are implemented in C++ to be able to access the native API of the different operating systems. The EA application has a modular structure so that new task types can be integrated later. This results in the software architecture shown in Fig. 6.2.

The entry point of the program is in the file *main.js*. This file loads the dependencies and the configuration of the EA application and initializes the GUI, which is realized as a web application. Additionally, it loads the student's certificate from the hard disk drive, which is used to authenticate against the EA server and to sign data stored on the EA server (*certificate.js*). Methods to sign data with a certificate and to verify signatures of other parties are available (*signature.js*). The EA application

---

[4]https://electronjs.org/.

[5]https://nodejs.org/en/.

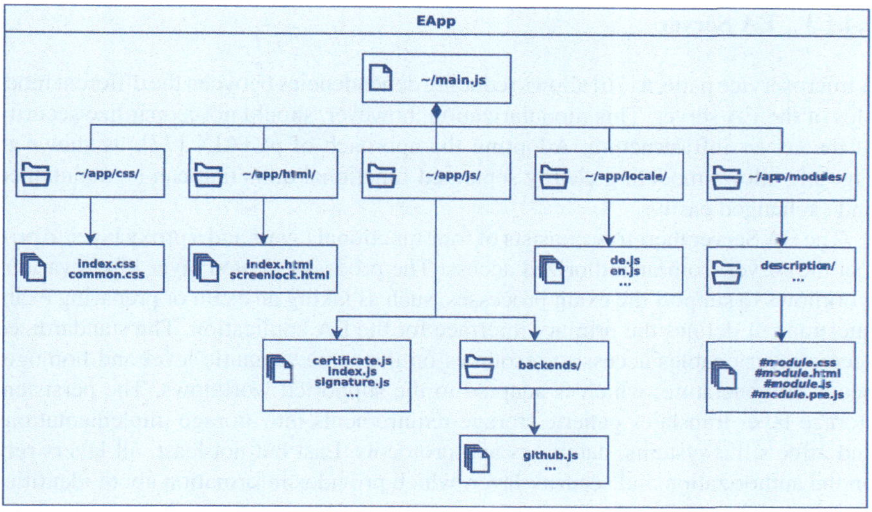

Fig. 6.2 Architecture of the EA application

provides two interfaces for extensions (*backends* and *modules*) as well as support for localization (*locale*).

The EA application is implemented to be fully portable. The design of the software works in such a way that running the EA application does not require administrative rights on the students' devices. This requirement is especially important because not all students may want to run a program that requires this type of privileges on their devices. In addition, the EA application does not use operating system storage mechanisms such as the *Registry* on a Windows system. All configurations are stored in a local configuration file. This allows the EA application to run on a system without interaction with the rest of the operating system, including the file system. For example, an additional user account that only has access to the EA application and the certificate could be created to prevent the EA application from being able to read other contents of the file system at all. Because the issue of privacy is very important, the source code of the EA application and the EA Server is published as an open source project. This allows students to check what happens with their certificate on their local computer and with their data on the EA Server before they agree to use the EA software.

The modular construction of the EA application allows to include the perceived advantages of EA, either raised by students or examiners, into the EA application. Basically, for every advantage a module can be integrated into the EA application. For example, the students mentioned *more realistic assignments* as an advantage of EA. Therefore, a module could introduce a type of assignment that fulfills this requirements. For instance, at a faculty for architecture a CAD tool could be integrated as a module, so that the students are able to use this tool from their later work life directly in the examination.

6.4.1.3 EA Server

A microservice pattern [76] allows reducing dependencies between the different modules of the EA server. This modularization, however, should not jeopardize security of the server infrastructure. Adopting the approach of pSTAIX [77], as shown in Fig. 6.3, allows modeling clearly separated functional units that can be maintained and exchanged easily.

The EA Server therefore consists of four functional layers and a proxy layer to protect the server from unauthorized access. The process-oriented layer offers various workflows to support the exam processes, such as taking an exam or preparing exam questions. It defines the primary interface for the EA application. The standardized access layer enables access to resources on the same semantic level and homogeneous nomenclature, which is adapted to the supported workflows. The persistent storage layer translates generic storage requirements into storage implementations and affects file systems, databases and protocols. Last but not least, all layers rely on the authorization and security layer, which provides information about identities and their roles within the processes as well as strong cryptography and signature functionalities to secure the workflows. If a microservice on any of the layers needs to write a logfile, this service can make use of the logging layer, which provides this functionality. In order to achieve a clear separation of concerns and to enable the reusability of the different modules in the layers, each module has clearly defined interfaces and dependencies. The levels are designed in such a way that higher levels may only depend on modules of the lower level, but not vice versa, in order to avoid circular dependencies.

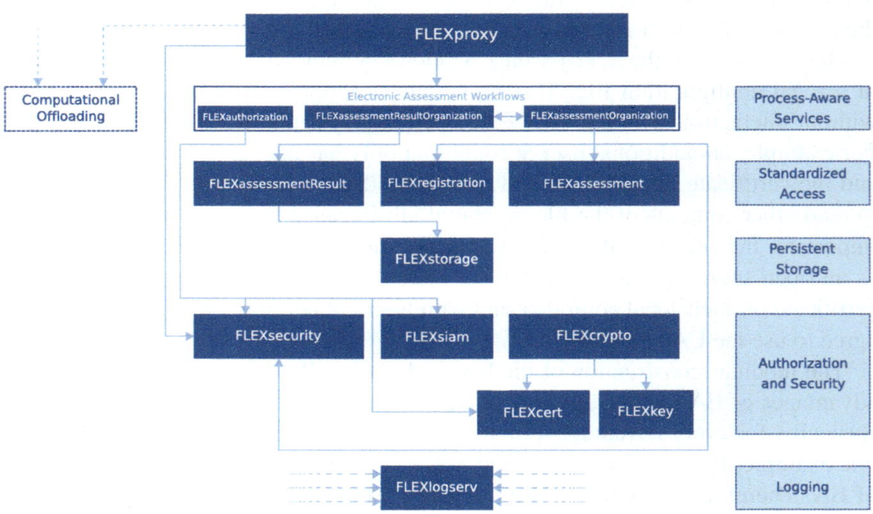

Fig. 6.3 Architecture of the EA server

All of the microservices provide a RESTful interface. The endpoints can potentially be used by students, examiners, and organizational staff. Every endpoint returns a status code *200* for a valid request, a status code *400* for a valid request with faulty parameters, or a status code *500* for an invalid or unauthorized request.

This server architecture allows also for the assessment of workgroups. This is possible due to the way the examination results are stored by the *FLEXstorage* micro service. It uses a *git*[6] backend. Therefore, the students' result are versioned and certain changes to the results can be traced back to a particular author. For individual examinations, the author will be the same for every update of the results on the server. However, for the assessment of a workgroup, the contributions to the final solution can be assigned to different students. That allows an examiner to determine an individual grade for each student, because the individual advocation into the solution can distinguished. Updates to the results that have been uploaded by one student can also be retrieved by the other students of a workgroup. Therefore, collaborative work on the assignments of an examination becomes possible. Additionally, storing the results in a versioned git repository on the server prevents data loss during the examination. The EA application implements a mechanism that automatically saves the students' results to the EA server in specified time intervals. If a student's device crashes during the examination, the results that have been created so far by that particular student can easily be retrieved from the server after the device has been restarted.

To prevent the manipulation of the examination results on the EA server, digital certificates are used to create a signature for each set of results. Therefore, the EA application uses a student's private key to create a signature of the results and uploads this signature alongside the results. This signature prevents the examiner from manipulating results. Additionally, the students do not have to be able to manipulate results. Therefore, the uploaded signature is signed with the EA server's private key. That means that both parties, the EA application and the EA server, have agreed upon the results that have been uploaded to the EA server. Therefore, neither the student nor the examiner can change an uploaded set of results, because in this case the signature of the other would be missing.

The possibility for *computational offloading* is important to smooth out differences between the students' devices. As will be discussed in Sect. 6.5, the performance of the EA application it dependent on the computer it is running on and on the architecture of the EA application itself. However, there may be computationally intensive tasks, for example executing a piece of source code, that may yield big differences in the performance on different devices. Therefore, these computationally intensive tasks can be offloaded to the EA server. In that way, all students have to rely on the same server to execute these tasks which can be considered fair. However, to make this approach work as intended, it is important to make the EA application available for at least all major operating systems. Otherwise, students would be handicapped if their operating system is not supported, which would not be fair.

---

[6]https://git-scm.com/.

To be able to execute the EA server, a physical or virtual server is needed. All the different microservices of the EA server are executed in docker containers in a virtual machine that has four virtual computing cores and eight gigabytes of RAM. However, the microservice for computational offloading was deployed to a physical server with 32 computing cores and 32 gigabytes of RAM, because the computational offloading is designed to carry out computationally intensive tasks. Therefore, it had to be expected that the smaller virtual machine that executes the other micro services would not scale to a large number of students that want to use computational offloading.

6.4.1.4 Examination Network

The connection between EA application and EA server is established via a special examination network (EN). This network is the only one through which the EA server can be accessed. This means in particular that a "regular" network of an HEI, e.g. *eduroam*, must not allow connections to the EA server. During an EA the user accounts for registered students are transferred from the regular network to the EN. Therefore, all registered students can only connect to the EN during the EA, not to other networks in the HEI. In addition, each Student can only use the EN credentials to establish one connection, thus multiple connections with the same credentials are not possible. The EN manages network traffic with a firewall and prevents any connections between two clients. However, it is not possible to prevent students from establishing point-to-point connections over ad-hoc Wi-Fi networks or even mobile internet connections such as LTE. Therefore, this is a threat that has to be resolved in the EA application.

For most HEIs that already provide a WiFi network, there is no additional hardware needed to deploy an EN. Rather, the already existing hardware can be used to generate a new WiFi network service set identifier (SSID) that fulfills the requirements of the EN.

6.4.1.5 Invigilator Tablet

The Invigilator Tablet (IT) is mainly used to replace paper-based registration lists. It can download the current registration list from the EA Server so that invigilators can use this digital list for student registration in a similar way as current paper-based examinations. Students can sign their participation in the exam on the tablet. In addition, the IT has the ability to verify a client's connection by scanning a QR code displayed in the EA application, which contains information about the student, such as full name, and the version of the EA application being executed. This feature is important to make it more difficult for students to cheat during an exam. In an EA this would be possible if another person, who is not present in the exam room, would log on to the EA Server with a particular student's credentials and certificate. However, that student would have to be present in the exam room to have his participation in the exam properly registered, and the EA application would have to look like an unmodified version properly logged in to the EA server to the invigilator.

Every recent tablet should be feasible to serve as the IT. As the application running on the IT does not perform computationally intensive tasks, but provides only a management interface that connects to the EA server, the requirements to the hardware are very low. However, it has to be able to connect to the examination network. Therefore, it is necessary to obey the hardware specifications of the examination network. However, these should be very easy to match with a recently bought device.

6.4.2 Organizational Framework

Not every requirement could be met by the software design and implementation. Therefore the fulfillment of certain requirements must be ensured by a set of rules and boundary conditions.

Students and examiners must be able to examine **FLEX** and all of its components. For students, the internal functions of the EA application are of particular importance, because this is the software they are supposed to run on their personal devices. For examiners, it is important to see if **FLEX** is capable of meeting their requirements for examinations.

All data, be it the students' results or the tasks of an exam, must be stored in a loss-proof manner. Since it cannot be guaranteed that a file storage system will work properly, this must be achieved by backups. For this purpose, a tape system [78] could be used, for example, as this kind of fulfils the very similar requirements of a research data management system [79].

As already discussed, it is not sufficient to issue certificates that are only valid within the scope of EA. Rather, the use of certificates to identify students in a digital process must be implemented for the entire (digital) life cycle of students.

Students in particular have called for more realistic examinations. The idea behind this demand was to include more of the skills taught in the courses into the examinations. In computer science, for example, this includes skills such as handling a debugger or interpreting error messages from the compiler.

Although most students have equipment that is suitable for performing EA, some students may not have access to a suitable device. In order to enable these students to participate in an EA, the HEI must provide rental equipment. These students, but also students who have their own device but do not have distinct computer skills may encounter problems when instructions refer to digital certificates and portable applications. Therefore, the HEI must provide technical support.

6.5 Evaluation and Testing

To be able to judge on the achieved results with **FLEX**, it is important to test and evaluate the resulting software prototype. The focus of the evaluation will be put on *usability*, *performance*, and *security*.

The *usability* of EA application was measured using the *System Usability Scale* (SUS [80, 81]), which is a questionnaire commonly used to assess the usability of a piece of software. The questionnaire for *SUS* consists of only ten Likert-items (q_1 to q_{10}), ranging from 1 to 5. It can be found in the appendix of this chapter. Five of these items formulate a positive statement, the other five items formulate a negative statement. Based on these questions, *SUS* assigns a score between 25 and 100 to each questionnaire using the following formula:

$$s = 2.5 * \sum_{i=1}^{10} q_i \tag{6.1}$$

To determine the usability of the EA application, an alpha test with 160 *MATSE* trainees was carried out. At the time of the test, 101 of these trainees were freshmen, the remaining 59 trainees were in their third semester of study. The students had so solve a simple programming task using the EA application. Afterwards, they were handed a copy of the *SUS* questionnaire which they were asked to fill out. The analysis of the results revealed a good to excellent usability for the EA application based on the usability classes that *SUS* assigns to particular scores. The mean value of all of the questionnaires is $\mu = 81.09375$. A distribution of the results is depicted in Fig. 6.4.

The *performance* of the EA applications is important for the equal treatment of the students. If the performance of the client software differs too much between devices with different capabilities, this would hinder students with slower devices. For each available test device and configuration, the EA application was started 100 times using a shell script. For each run, a file was fetched from the EA server, automatically modified, and saved back to the EA server. Timings were measured

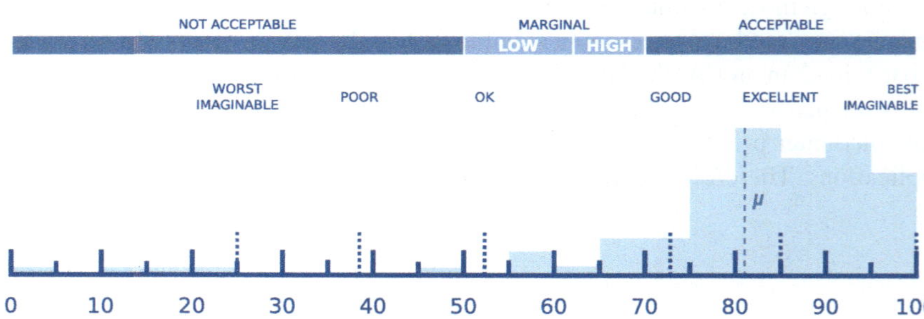

Fig. 6.4 System usability scale: results

for the whole execution cycle, as well as for the single steps. The achieved results are very promising regarding the performance of the EA application on different devices. First of all, the results show that a better equipped device does not necessarily lead to a better performance. Second, the results allow for the conclusion that the EA application allows for the equal treatment of the students during the examination. The performance gap between the best and worst test device was about one second. This can be considered negligible as even differences in the speed of the students' handwriting are considered negligible, even if those can have a way greater effect.

Security can be considered the most important requirement for a software framework for EA. At the same time, it is also the requirement that is most difficult to prove that compliance for. In order to take this into account, a threat model [82] was developed, which takes into account both, requirements explicitly formulated by students, examiners and administrative bodies and implicit requirements from general security concepts. For all the identified requirements a technical or organizational measure could be found to fulfill this specific requirement. The publication of the EA framework as open source software additionally contributes to the security of the software [83] and ensures that missing requirements and potentially existing errors in the implementation can be identified and corrected by a community software review.

6.6 Summary and Outlook

Starting from the question whether it is possible to develop an application for electronic exams on students' devices in a bring-your-own-device environment, the project that led to the development of **FLEX** was launched. The main goal was to have a software solution available that enables EA within the MATSE program, where a BYOD approach is the only possible solution.

6.6.1 Challenges and Lessons Learned

When the project started, we faced several challenges. First of all, it was a very tedious (and actually non-deterministic) task to find out about the boundary conditions of EA in a BYOD scenario that are given by laws and regulations. As it turned out, EA, at least in Germany, is an area which is regulated by laws, but which are rarely applied in legal cases. Therefore, no judgments establishing a principle are available as a guide line. However, asking lawyers does not clarify the boundary conditions either. The answers that we received were rather vague and sometimes even contradicted each other. Thus, we decided to take the documents available at RWTH Aachen University as our guide line for the boundary conditions the EA framework had to be developed in. In general, for us it seemed that not asking too many questions worked much better than trying to clear up any doubt before starting the project.

Once the legal boundary conditions were cleared, other challenges had to be tackled. As the next step in the research process, the requirements engineering posed another challenge. The design of the questionnaire and the workflow for its analysis were actually the easy part. The hard part was reaching enough people to get a decent data set to draw reliable conclusions from. Having visited a lot of conferences and workshops over the last years proved to be very valuable here, as the contacts made there worked as multipliers for spreading the online questionnaire.

Last, but not least, technical challenges had to be faced. Solutions had to be found for a software architecture that allowed for the equal treatment of students, including multi-platform support and computational offloading, and for the verification of an unmodified EA application on untrusted devices. Do accomplish this, we had to bring technologies from different fields of computing, e.g. mobile computer and security engineering, together into one software in order to make things work.

6.6.2 Future Work

Even if a functional prototype is available that meets the collected requirements, the project is not finished. With a flexible software solution at hand, the focus shifts to the practical application of the software. Therefore a field test is planned for September 2020 within the framework of the MATSE educational program. It was decided that the first Java exam of the freshmen should be written with **FLEX**. This means that about 200 trainees will use **FLEX** for the first exam *in the wild*.

To make the use of **FLEX** attractive for other HEIs as well, it proved to be important to enable other tools with **FLEX**, e.g. an LMS for the creation of exams. In addition, support for a wider range of mobile devices, such as tablet computers and perhaps even smartphones, is planned. It is undeniable that these devices are not suitable for every type of exam for which a notebook is suitable; however, there are types of tasks, e.g. multiple choice tasks, that can be solved with such devices. When porting **FLEX** to mobile operating systems like Android and iOS, it is clear in advance that not all implemented anti-cheating measures will work on these platforms. Therefore a substantial amount of work will have to be invested to solve the special requirements of anti-cheating measures on mobile platforms.

Beyond the practical application and with a much stronger focus on research, the possibilities that come with a software like **FLEX** are evaluated. Of particular interest is the automatic correction of programming exercises and the integration of such mechanisms in **FLEX**. Furthermore, new types of assignments are of interest. For a software developer it is actually not a common scenario to develop a very limited task on his own. Rather, the development of more complex tasks in a development team is the reality at the workplace. Therefore, integrating teamwork into an exam while retaining the ability to give students individual grades is another interesting topic in the MATSE context. A third topic of interest is *accessibility*. Technology is one way of overcoming certain limitations of a disability in general [84] and the same is true for the field of education [85]. Therefore, a framework for EA opens the possibility to increase the ability to take exams for disabled students in a way that a paper-based exam never could.

Appendix

Surveys

Students

The particular options for the tagged items as shown in Fig. 6.5 are:

1. <18 Years; 18-25 Years; >25 Years
2. BSc. Computer Science, MSc. Computer Science, BSc. Scientific Programming / AMI, MSc. Technomathematics, BSc. Technical Communication, MSc. Technical Communication, BSc. Computer Science (Teacher), MSc. Computer Science (Teacher), Other (free text)
3. Faster Correction, More Realistic Examinations, More Diverse Examination Tasks, Other (free text)
4. Security, Usability, Fairness, Other (free text)
5. Familiar Device, Location-independent Examinations, Other (free text)
6. Security, Differences Between Devices, Other (free text).

Examiners

The particular options for the tagged items as shown in Fig. 6.6 are:

1. <30 Years; 30–50 Years; >50 Years
2. Humanities

   - Archaeology, Ethics, History, Cultural Studies, Literature Studies, Philosophy, Theology, Linguistics, Other Humanities

   Engineering

   - Civil Engineering, Biotechnology, Electrical Engineering, Information Technology, Mechanical Engineering, Medical Engineering, Environmental Engineering, Process Engineering, Materials Engineering, Other Engineering Science

Medical Sciences

- Health Sciences, Human Medicine, Pharmaceutics, Veterinary Medicine, Dental Medicine, Other Medical Science

Natural Sciences

- Biology, Chemistry, Geoscience, Computer Sciences, Mathematics, Physics, Other Natural Science

Social Sciences

- Comparative Education, Human Geography, Communication Studies, Media Studies, Political Sciences, Psychology, Laws, Sociology, Economics, Other Social Science

3. Diverse, Female, Male
4. 1–5 Semesters; 6–10 Semesters; 10–20 Semesters; >20 Semesters;
5. Faster Correction, More Realistic Examinations, More Diverse Examination Tasks, Other (free text)
6. Security, Usability, Fairness, Uncertain Legal Situation, Other (free text)
7. Familiar Device, Location-independent Examinations, Cost Reduction for the HEI, Other (free text)
8. Security, Differences Between Devices, Other (free text).

Students

Part	Item	Scale
General	Age	3 Options[1]
	Study Program	9 Options[2]
	Gender	2 Options
TA-EG	I like to have new electronic devices.	Five-level Likert Scale (5LLS)
	Electronic devices cause illness.	
	I like to go to stores for electronic devices.	
	I (would) have problems understanding electronic and computer magazines	
	Electronic devices provide a high standard of living.	
	Electronic devices lead to intellectual impoverishment.	
	Electronic devices make many things more complicated.	
	I inform myself about electronic devices, even if I have no intention to buy them.	
	Electronic devices make you independent.	
	I enjoy trying out electronic devices.	
	Electronic devices make everyday life easier for me.	
	Electronic devices increase security.	
	Electronic devices reduce personal contact between people.	
	I know most of the functions of the electronic devices I own.	
	I am thrilled when a new electronic device comes onto the market.	
	Electronic devices cause stress.	
	I know about electronic devices.	
	It is easy for me to learn how to operate an electronic device.	
	Electronic devices help to obtain information.	
e-Assessment	E1: I think it is very good to have electronic examinations in my studies.	5LLS
	E2: I think that electronic examinations are a good complement to paper-based examinations.	
	E3: I think that electronic examinations are a good substitute to paper-based examinations.	
	E4: I see advantages of electronic examinations, namely...	4 Options[3]
	E5: I see disadvantages of electronic examinations, namely...	4 Options[4]
BYOD	B1: I find it very advantageous if electronic examinations are carried out on my own electronic device (laptop).	5LLS
	B2: I see the following advantages in using my own electronic device (laptop) for an examination, namely...	3 Options[5]
	B3: I see the following disadvantages in using my own electronic device (laptop) for an examination, namely....	3 Options[6]
Fraud	C1: I think that cheating in paper-based examinations can be done very easy.	5LLS
	C2: I think that cheating in electronic examinations can be done very easy.	

Fig. 6.5 Students' survey (originally in German)

Part	Item	Scale
General	Age	3 Options[1]
	Field of Expertise	47 Options[2]
	Gender	3 Options[3]
	Teaching Experience	4 Options[4]
TA-EG	I like to have new electronic devices.	Five-level Likert Scale (5LLS)
	Electronic devices cause illness.	
	I like to go to stores for electronic devices.	
	I (would) have problems understanding electronic and computer magazines.	
	Electronic devices provide a high standard of living.	
	Electronic devices lead to intellectual impoverishment.	
	Electronic devices make many things more complicated.	
	I inform myself about electronic devices, even if I have no intention to buy them.	
	Electronic devices make you independent.	
	I enjoy trying out electronic devices.	
	Electronic devices make everyday life easier for me.	
	Electronic devices increase security.	
	Electronic devices reduce personal contact between people.	
	I know most of the functions of the electronic devices I own.	
	I am thrilled when a new electronic device comes onto the market.	
	Electronic devices cause stress.	
	I know about electronic devices.	
	It is easy for me to learn how to operate an electronic device.	
	Electronic devices help to obtain information.	
e-Assessment	E1: I think it is very good to have electronic examinations in my studies.	5LLS
	E2: I think that electronic examinations are a good complement to paper-based examinations.	
	E3: I think that electronic examinations are a good substitute to paper-based examinations.	
	E4: I see advantages of electronic examinations, namely…	4 Options[5]
	E5: I see disadvantages of electronic examinations, namely…	4 Options[6]
BYOD	B1: I find it very advantageous if electronic examinations are carried out on my own electronic device (laptop).	5LLS
	B2: I see the following advantages in using my own electronic device (laptop) for an examination, namely...	3 Options[7]
	B3: I see the following disadvantages in using my own electronic device (laptop) for an examination, namely....	3 Options[8]
Fraud	C1: I think that cheating in paper-based examinations can be done very easy.	5LLS
	C2: I think that cheating in electronic examinations can be done very easy.	

Fig. 6.6 Teachers' survey (originally in German)

Software Usability Scale

See Fig. 6.7.

	1	2	3	4	5
I think that I would like to use this system frequently.	☐	☐	☐	☐	☐
I found the system unnecessarily complex.	☐	☐	☐	☐	☐
I thought the system was easy to use.	☐	☐	☐	☐	☐
I think that I would need the support of a technical person to be able to use this system.	☐	☐	☐	☐	☐
I found the various functions in this system were well integrated.	☐	☐	☐	☐	☐
I thought there was too much inconsistency in this system.	☐	☐	☐	☐	☐
I would imagine that most people would learn to use this system very quickly.	☐	☐	☐	☐	☐
I found the system very cumbersome to use.	☐	☐	☐	☐	☐
I felt very confident using the system.	☐	☐	☐	☐	☐
I needed to learn a lot of things before I could get going with this system.	☐	☐	☐	☐	☐

Fig. 6.7 Software usability scale (originally in German)

References

1. Politze, M., Schaffert, S., & Decker, B. (2016). A secure infrastructure for mobile blended learning applications. In *Proceedings EUNIS 2016 , Ser. EUNIS Proceedings* (Vol. 22, pp. 49–56). http://www.eunis.org/download/2016/EUNIS2016_paper_19.pdf. Accessed January 13, 2020.
2. Digitalisierung, H. (2016). The digital turn: Hochschulbildung im digitalen Zeitalter, Geschäftsstelle Hochschulforum Digitalisierung beim Stifterverband für die Deutsche Wissenschaft e.V. (Ed.), Berlin. https://hochschulforumdigitalisierung.de/sites/default/files/dateien/Abschlussbericht.pdf. Accessed January 13, 2020.
3. Walker, R., & Handley, Z. (2016). Designing for learner engagement with computer-based testing. *Research in Learning Technology, 24*, 88. ISSN 2156-7077. https://doi.org/10.3402/rlt.v24.30083.
4. Terzis, V., & Economides, A. A. (2011). The acceptance and use of computer based assessment. *Computers & Education, 56*(4), 1032–1044. ISSN 0360-1315. https://doi.org/10.1016/j.compedu.2010.11.017.
5. Luecht, R. M., & Sireci, S. G. (2011). *A review of models for computer-based testing: research report 2011–2012*. https://www.researchgate.net/publication/265622331_A_Review_of_Models_for_Computer-Based_Testing. Accessed January 13, 2020.
6. Birch, D., & Burnett, B. (2009). Bringing academics on board: Encouraging institution-wide diffusion of e-learning environments. *Australasian Journal of Educational Technology, 25*(1), ISSN 1449-5554. https://doi.org/10.14742/ajet.1184.
7. Vogt, M., & Schneider, S. (2009). E-Klausuren an Hochschulen: Didaktik - Technik - Systeme - Recht - Praxis. In Justus-Liebig-Universität Gießen (Ed.), Gießen. http://geb.uni-giessen.de/geb/volltexte/2009/6890/. Accessed January 13, 2020.
8. Biella, D., Engert, S., & Huth, D. (2009). Design and delivery of an e-assessment solution at the University of Duisburg-Essen. In *Proceedings EUNIS 2009, Ser. EUNIS Proceedings* (Vol. 15). https://www.uni-due.de/imperia/md/content/zim/veranstaltungen/eunis_09_e-assessment.pdf. Accessed January 13, 2020.
9. Bücking, J. (2010). eKlausuren im Testcenter der Universität Bremen: Ein Praxisbericht. https://www.campussource.de/events/e1010tudortmund/docs/Buecking.pdf. Accessed January 13, 2020.
10. Dahlstrom, E., Brooks, C., Grajek, S., & Reeves, J. (2015). *Undergraduate students and IT*. Louisville. https://library.educause.edu/%7E/media/files/library/2015/8/ers1510ss.pdf?la=en. Accessed January 13, 2020.
11. Poll, H. (2015). *Student mobile device survey 2015: National report: College students*. http://www.pearsoned.com/wp-content/uploads/2015-Pearson-Student-Mobile-Device-Survey-College.pdf. Accessed January 13, 2020.
12. Willige, J. (2016). Auslandsmobilität und digitale Medien: Arbeitspapier Nr. 23, Geschäftsstelle Hochschulforum Digitalisierung beim Stifterverband für die Deutsche Wissenschaft e.V (Ed.), Berlin. https://hochschulforumdigitalisierung.de/de/arbeitspapier-auslandsmobilitaet-digitale-medien. Accessed January 13, 2020.
13. Johnson, L., Becker, S. A., Estrada, V., & Freeman, A. (2015). NMC Horizon Report: 2016 Higher Education Edition. *Technical Report*. Austin, Texas: The New Media Consortium. https://library.educause.edu/-/media/files/library/2015/2/hr2015-pdf.pdf. Accessed January 13, 2020.
14. Johnson, L., Becker, S. A., Cummins, M., Estrada, V., Freeman, A., & Hall, C. (2016). NMC Horizon Report: 2016 Higher Education Edition. *Technical Report*. Austin, Texas: The New Media Consortium. https://library.educause.edu/-/media/files/library/2016/2/hr2016.pdf. Accessed January 13, 2020.

15. Küppers, B., Eifert, T., Politze, M., & Schroeder, U. (2018). E-assessment behind the scenes, common perceptions of e-assessment and how we see it nowadays. In *Proceedings of the 10th International Conference on Computer Supported Education—Vol. 2, CSEDU 2018—10th International Conference on Computer Supported Education, Funchal, Madeira (Portugal), 15 Mar 2018–17 Mar 2018* (pp. 285–292). SciTe Press. ISBN 978-989-758-291-2. https://doi.org/10.5220/0006788402850291.
16. Wang, F., & Hannafin, M. (2005). Design-based research and technology-enhanced learning environments. *Educational Technology Research and Development, 53*, 5–23. https://doi.org/10.1007/BF02504682.
17. March, S. T., & Smith, G. F. (1995). Design and natural science research on information technology. *Decision Support Systems, 15*(4), 251–266. https://doi.org/10.1016/0167-9236(94)00041-2.
18. Hevner, A. R., March, S. T., Park, J., & Ram, S. (2004). Design science in information systems research. *MIS Quarterly, 28*(1), 75–105. ISSN 0276-7783. http://www.jstor.org/stable/25148625. Accessed January 13, 2020.
19. Fluck, A., & Hillier, M. (2017). eExams: Strength in diversity. In *IFIP advances in information and communication technology* (pp. 409–417). Springer. https://doi.org/10.1007/978-3-319-74310-3_42.
20. Melve. I. (2013). BYOD for exams: leaving students to their own devices. https://de.slideshare.net/imelve/tnc-melve20130605v04. Accessed January 14, 2020.
21. Küppers, B., & Schroeder, U. (2016). Bring your own device for e-assessment, a review. In L. Gòmez Chova, A. Lòpez Martìnez, & I. Candel Torres (Eds.), *EDULEARN 2016 : 8th International Conference on Education and New Learning Technologies, Barcelona (Spain), 4 Jul 2016–6 Jul 2016* (pp. 8770–8776). Valencia: IATED Academy. ISBN 978-84-608-8860-4. https://doi.org/10.21125/edulearn.2016.0919.
22. Küppers, B., & Schroeder, U. (2018). A framework for e-assessment on students' devices, technical considerations. In E. Ras, & A. E. Guerrero Roldàn (Eds.), *Technology Enhanced Assessment, ser. Communications in Computer and Information Science, TEA 2017—20th International Conference, Barcelona (Spain), 5 Oct–6 Oct 2017* (Vol. 829, pp. 83–95). Cham: Springer. ISBN 978-3-319-97807-9. https://doi.org/10.1007/978-3-319-97807-9_7.
23. Lukarov, V. (2019). *Scaling up learning analytics in blended learning Szenarien* (Dissertation). Aachen: RWTH Aachen University. https://doi.org/10.18154/rwth-2019-05165
24. Küppers, B., Dondorf, T., Willemsen, B., Pflug, H. J., Vonhasselt, C., Magrean, B., et al. (2016). The scientific programming integrated degree program—A pioneering approach to join theory and practice. *Procedia Computer Science, 80*, 1957–1967. https://doi.org/10.1016/j.procs.2016.05.516.
25. Nacken, H., & Knight, C. (2019). *Digitalization strategy for teaching*. http://www.rwth-aachen.de/cms/%7Ehjfu/?lidx=1. Accessed January 14, 2020.
26. RWTH Aachen University. (2018). Digitalisierungsstrategie der Lehre an der RWTH Aachen - Die zweite Phase 2018–2023. http://www.rwth-aachen.de/global/show_document.asp?id=aaaaaaaaaayitvk. Accessed January 14, 2020.
27. Glaser, S. (2015). Ordnung zum Schutz personenbezogener Daten bei multimedialer Nutzung von E-Learning-Verfahren an der Rheinisch-Westfälischen Technischen Hochschule Aachen. http://www.rwth-aachen.de/global/show_document.asp?id=aaaaaaaaaaolwek. Accessed January 14, 2020.
28. Hillier, M. (2015). e-exams with student owned devices: Student voices. In *Proceedings of the International Mobile Learning Festival 2015* (pp. 582–608). http://transformingexams.com/files/Hillier_IMLF2015_full_paper_formatting_fixed.pdf. Accessed January 14, 2020.
29. Jawaid, M., Moosa, F. A , Jaleel, F., & Ashraf, J. (2014). Computer based assessment (CBA): Perception of residents at Dow University of Health Sciences. *Pakistan Journal of Medical Sciences, 30*(4), 688–691. https://doi.org/10.12669/pjms.304.5444.
30. Alsadoon, H. (2017). Students' perceptions of e-assessment at Saudi Electronic University. *The Turkish Online Journal of Educational Technology, 16*(1), 147–153. https://eric.ed.gov/?id=EJ1124924. Accessed January 14, 2020.

31. Babo, R. B., Azevedo, A. I., & Suhonen, J. (2015). Students' perceptions about assessment using an e-learning platform. In *2015 IEEE 15th International Conference on Advanced Learning Technologies* (pp. 244–246). https://doi.org/10.1109/ICALT.2015.73.
32. Sorensen, E. (2013). Implementation and student perceptions of e-assessment in a chemical engineering module. *European Journal of Engineering Education*, *38*(2), 172–185. https://doi.org/10.1080/03043797.2012.760533.
33. Hodgson, P., & Pang, M. Y. (2012). Effective formative e-assessment of student learning: A study on a statistics course. *Assessment & Evaluation in Higher Education*, *37*(2), 215–225. https://doi.org/10.1080/02602938.2010.523818.
34. Özden, M., Etürk, I., & Sanli, R. (2004). Students' perceptions of online assessment: A case study. *Journal of Distance Education*, *19*(2), 77–92. https://eric.ed.gov/?id=EJ807820. Accessed January 14, 2020.
35. Dermo, J. (2009). e-assessment and the student learning experience: A survey of student perceptions of e-assessment. *British Journal of Educational Technology*, *40*(2), 203–214. https://doi.org/10.1111/j.1467-8535.2008.00915.x.
36. Deutsch, T., Herrmann, K., Frese, T., & Sandholzer, H. (2012). Implementing computer-based assessment—A web-based mock examination changes attitudes. *Computers & Education*, *58*(4), 1068–1075. ISSN 0360-1315. https://doi.org/10.1016/j.compedu.2011.11.013.
37. Jimoh, R. G., Shittu, A. K., & Kawu, Y. K. (2012).Students' perception of computer based test (CBT) for examining undergraduate chemistry courses. *Journal of Emerging Trends in Computing and Information Sciences*, *3* (2), 125–134. ISSN 2079-8407. http://www.cisjournal.org/journalofcomputing/archive/vol3no2/vol3no2_2.pdf. Accessed January 14, 2020.
38. Nardi, A., & Ranieri, M. (2018). Comparing paper-based and electronic multiple-choice examinations with personal devices: Impact on students' performance, self-efficacy and satisfaction. *British Journal of Educational Technology*,. https://doi.org/10.1111/bjet.12644.
39. Hochlehnert, A., Brass, K., Moeltner, A., & Juenger, J. (2011). Does medical students' preference of test format (computer-based vs. paper-based) have an influence on performance? *BMC Medical Education*, *11*(1).https://doi.org/10.1186/1472-6920-11-89.
40. Nikou, S. A., & Economides, A. A. (2016). The impact of paper-based, computer-based and mobile-based self-assessment on students' science motivation and achievement. *Computers in Human Behavior*, *55*, 1241–1248. https://doi.org/10.1016/j.chb.2015.09.025.
41. Williams, B. (2007). Students' perceptions of prehospital web-based examinations. *International Journal of Education and Development using Information and Communication Technology*, *3*(1), 54–63. https://www.learntechlib.org/d/42354. Accessed January 14, 2020.
42. Ogilvie, R. W., Trusk, T. C., & Blue, A. V. (1999). Students' attitudes towards computer testing in a basic science course. *Medical Education*, *33*(11), 828–831. https://doi.org/10.1046/j.1365-2923.1999.00517.x.
43. Lim, E. C., Ong, B. K., Wilder-Smith, E. P., & Seet, R. C. (2006). Computer-based versus pen-and-paper testing: students' perception. *Annals, Academy of Medicine, Singapore*, *35*(9), 599–603. http://www.annals.edu.sg/pdf/35VolNo9Sep2006/V35N9p599.pdf. Accessed January 14, 2020.
44. Boeve, A. J., Meijer, R. R., Albers, C. J., Beetsma, Y., & Bosker, R. J. (2015). Introducing computer-based testing in high-stakes exams in higher education: results of a field experiment. *PloS one*, *10*(12). ISSN 1932-6203. https://doi.org/10.1371/journal.pone.0143616.
45. K. Pforr, Dannwolf, T. (2017). What do we lose with online-only surveys? Estimating the bias in selected political variables due to online mode restriction. *Statistics, Politics and Policy*, *8*(1). https://doi.org/10.1515/spp-2016-0004.
46. Karrer, K., Glaser, C., Clemens, C., & Bruder, C. (2009). Technikaffinität erfassen – der Fragebogen TA-EG. In A. Lichtenstein, C. Stößel, & C. Clemens (Eds.), *Der Mensch im Mittelpunkt technischer Systeme. 8. Berliner Werkstatt Mensch-Maschine-Systeme, ser. ZMMS Spektrum* (Vol. 22, pp. 196–201). Düsseldorf, Germany: VDI Gerlag GmbH.
47. Kaufmann, L. (1987). Clustering by means of medoids. In *Proceedings Statistical Data Analysis Based on the L1 Norm Conference, Neuchatel* (pp. 405–416)

48. Sheskin, D. J. (2003). *Handbook of parametric and nonparametric statistical procedures* (3rd ed.). CRC Press. ISBN 9781420036268.
49. Prensky, M. (2001). Digital natives, digital immigrants Part 1. *On the Horizon*, *9*(5), 1–6. https://doi.org/10.1108/10748120110424816.
50. Turner, A. (2015). Generation Z: Technology and social interest. *The Journal of Individual Psychology*, *71*(2), 103–113. https://doi.org/10.1353/jip.2015.0021.
51. Kadijevich, D. (2000). Gender differences in computer attitude among ninth-grade students. *Journal of Educational Computing Research*, *22*(2), 145–154. https://doi.org/10.2190/k4u2-pwqg-re8l-uv90.
52. Kahveci, M. (2010). Students perceptions to use technology for learning: measurement Integrity of the modified Fennema-Sherman attitudes scales. *Turkish Online Journal of Educational Technology—TOJET*, *9*(1), 185–201.
53. Yau, H.K., & Cheng, A. L. F. (2012). Gender difference of confidence in using technology for learning. *The Journal of Technology Studies*, *38*(2). https://doi.org/10.21061/jots.v38i2.a.2.
54. Werner, P., & Karnieli, E. (2003). A model of the willingness to use telemedicine for routine and specialized care. *Journal of Telemedicine and Telecare*, *9*(5), 264–272. https://doi.org/10.1258/135763303769211274.
55. Kang, M., & Gretzel, U. (2012). Perceptions of museum podcast tours: Effects of consumer innovativeness, internet familiarity and podcasting affinity on performance expectancies. *Tourism Management Perspectives*, *4*, 155–163. https://doi.org/10.1016/j.tmp.2012.08.007.
56. Kocdar, S., Karadeniz, A., Peytcheva-Forsyth, R., & Stoeva, V. (2018). Cheating and plagiarism in e-assessment: Students' perspectives. *Open Praxis*, *10*(3), 221. https://doi.org/10.5944/openpraxis.10.3.873.
57. Hillier, M., Grant, S., & Coleman, M. A. (2018). Towards authentic e-exams at scale: Robust networked moodle. In M. Campbell, J. Willems, C. Adachi, D. Blake, I. Doherty, S. Krishnan, et al. (Eds.), *ASCILITE 2018–Conference Proceedings , Geelong (Australia), 25 Nov 2018–28 Nov 2018* (Vol. 35, pp. 131–141). http://ascilite.org/wp-content/uploads/2018/12/ASCILITE-2018-Proceedings.pdf. Accessed January 14, 2020.
58. Küppers, B., & Schroeder, U. (2019). Students' perceptions of e-assessment. In D. Passey, R. Bottino, C. Lewin, & E. Sanchez (Eds.), *IFIP Advances in Information and Communication Technology, Ser. IFIP Advances in Information and Communication Technology. IFIP TC 3 Open Conference on Computers in Education (OCCE) 2018, Linz (Austria), 24 Jun 2018–28 Jun 2018* (Vol. 524, pp. 275–284). Cham: Springer. ISBN 978-3-030-23513-0. https://doi.org/10.1007/978-3-030-23513-0_27.
59. Rolim, C., & Isaias, P. (2018). Examining the use of e-assessment in higher education: Teachers and students' viewpoints. *British Journal of Educational Technology*, *4*, 1785–1800. https://doi.org/10.1111/bjet.12669.
60. Mellar, H., Peytcheva-Forsyth, R., Kocdar, S., Karadeniz, A., & Yovkova, B. (2018). Addressing cheating in e-assessment using student authentication and authorship checking systems: Teachers' perspectives. *International Journal for Educational Integrity*, *14*(1), 2. ISSN 1833-2595. https://doi.org/10.1007/s40979-018-0025-x.
61. Freiwald, T., Salimi, M., Khaljani, E., & Harendza, S. (2014). Pattern recognition as a concept for multiple-choice questions in a national licensing exam. *BMC Medical Education*, *14*(1). https://doi.org/10.1186/1472-6920-14-232.
62. Jamil, M., Tariq, R. H., & Shami, P. A. (2012). Students perceptions to use technology for learning: Measurement Integrity of the modified Fennema-Sherman attitudes acales. *Turkish Online Journal of Educational Technology—TOJET*, *11*(4), 371–381.
63. Kuikka, M., Kitola, M., & Laakso, M.-J. (2014). Challenges when introducing electronic exam. *Research in Learning Technology*, *22*. https://doi.org/0.3402/rlt.v22.22817.
64. Küppers, B., & Schroeder, U. (2020). Teacher's perspective on e-assessment. In *Proceedings of the 12th International Conference on Computer Supported Education—Vol. 1: CSEDU, CSEDU 2020—12th International Conference on Computer Supported Education, Online Streaming, 02 May 2020–04 May 2020* (pp. 495–502). SciTe Press. ISBN 978-989-758-417-6. https://doi.org/10.5220/0009578004950502.

65. Sheard, J., Markham, S., & Dick, M. (2003). Investigating differences in cheating behaviours of IT undergraduate and graduate students: The maturity and motivation factors. *Higher Education Research & Development*, *22*(1), 91–108. https://doi.org/10.1080/0729436032000056526.
66. Schneier, B. (2008). The psychology of security. In S. Vaudenay (Ed.), *Progress in Cryptology—AFRICA-CRYPT 2008* (pp. 50–79). Berlin, Heidelberg: Springer. ISBN 978-3-540-68164-9. https://www.schneier.com/academic/paperfiles/paper-psychology-of-security.pdf. Accessed January 15, 2020.
67. B. Schneier. (2008). The difference between feeling and reality in security. https://www.schneier.com/essays/archives/2008/04/the_difference_betwe.html. Accessed January 15, 2020.
68. Sindre, G., Vegendla, A. (2015). E-exams versus paper exams: A comparative analysis of cheating-related security threats and countermeasures. http://ojs.bibsys.no/index.php/NISK/article/view/298. Accessed January 15, 2020.
69. Akherfi, K., Gerndt, M., & Harroud, H. (2018). Mobile cloud computing for computation offloading: Issues and challenges. *Applied Computing and Informatics*, *14*(1), 1–16. https://doi.org/10.1016/j.aci.2016.11.002.
70. Kovachev, D., & Klamma, R. (2012). Framework for computation offloading in mobile cloud computing. *International Journal of Interactive Multimedia and Artificial Intelligence*, *1*(7), 6. https://doi.org/10.9781/ijimai.2012.171.
71. Seshadri, A., Luk, M., Perrig, A., van Doom, L., & Khosla, P. (2007). Pioneer: Verifying code integrity and enforcing untampered code execution on legacy systems. In *Advances in Information Security* (pp. 253–289). US: Springer. https://doi.org/10.1007/978-0-387-44599-1_12.
72. Garay, J. A., & Huelsbergen, L. (2006). Software integrity protection using timed executable agents. In *Proceedings of the 2006 ACM Symposium on Information, Computer and Communications Security—ASIACCS '06*. ACM Press. https://doi.org/10.1145/1128817.1128847.
73. Barak, B., Goldreich, O., Impagliazzo, R., Rudich, S., Sahai, A., Vadhan, S., et al. (2001). In *Advances in cryptology—CRYPTO 2001* (pp. 1–18). Springer, Berlin, Heidelberg. https://doi.org/10.1007/3-540-44647-8_1.
74. Collberg, C., Thomborson, C., & Low, D. (1997). A taxonomy of obfuscating transformations. Technical Report 148. Department of Computer Sciences, The University of Auckland. https://researchspace.auckland.ac.nz/bitstream/handle/2292/3491/TR148.pdf.
75. Collberg, C., & Thomborson, C. (2002). Watermarking, tamper-proofing, and obfuscation—Tools for software protection. *IEEE Transactions on Software Engineering*, *28*(8), 735–746. https://doi.org/10.1109/tse.2002.1027797.
76. Namiot, D., & Sneps-Sneppe, M. (2014). On micro-services architecture. *International Journal of Open Information Technologies*, *2*(9), 24–27.
77. Politze, M., Decker, B., & Eifert, T. (2017). Pstaix—A process-aware architecture to support research processes. In M. Eibl & M. Gaedke (Eds.), *INFORMATIK 2017* (pp. 1369–1380). Bonn: Gesellschaft für Informatik. https://doi.org/10.18420/in2017_137.
78. Stanek, D., Eifert, T. (2012). Maßnahmen für verlässliche und schnelle datenwiederherstellung. *PIK—Praxis der Informationsverarbeitung und Kommunikation* (Vol. 35, No. 3). https://doi.org/10.1515/pik-2012-0032.
79. Eifert, T., Schilling, U., Bauer, H.-J., Krämer, F., & Lopez, A. (2017). Infrastructure for research data management as a cross-university project. In *Human Interface and the Management of Information: Supporting Learning, Decision-Making and Collaboration* (pp. 493–502). Springer. https://doi.org/10.1007/978-3-319-58524-6_39.
80. Brooke, J. (2013). SUS: A retrospective. *Journal of usability studies*, *8*(2), 29–40.
81. Brooke, J. (1996). SUS—A quick and dirty usability scale. *Usability evaluation in industry*, *189*(194), 4–7.
82. Myagmar, S., Lee, A. J., & Yurcik, W. (2005). Threat modeling as a basis for security requirements. In *Proceedings of the IEEE Symposium on Requirements Engineering for Information Security*.

83. Hoepman, J.-H., Jacobs, B. (2007). Increased security through open source. *Communications of the ACM*, *50*(1), 79–83. https://doi.org/10.1145/1188913.1188921.
84. Alper, S., & Raharinirina, S. (2006). Assistive technology for individuals with disabilities: A review and synthesis of the literature. *Journal of Special Education Technology*, *21*(2), 47–64. https://doi.org/10.1177/016264340602100204.
85. Edyburn, D. L. (2017). Assistive technology and students with mild disabilities. *Focus on Exceptional Children*, *32*(9). https://doi.org/10.17161/fec.v32i9.6776.

Chapter 7
Antares: A Flexible Assessment Framework for Exploratory Immersive Environments

Joachim Maderer and Christian Gütl

Abstract Despite several differences in contemporary learning theories and approaches to education, there is strong agreement that scaffolding and feedback are essential to foster successful learning processes. Technological advances have enabled the development of smart learning environments, e.g., intelligent tutoring systems, which provide authoring features for learning material with encapsulated assignments and customized automated guidance that directs learners to the current solution. In the field of STEM education, exploratory learning strategies are considered particularly effective. These approaches allow learners to construct their own knowledge—either individually or as a social group—by performing experiments or using engaging immersive 3D computer simulations. While basic simulations and virtual laboratories are widely accepted and used by physics teachers, most tools are very limited in terms of educational features such as internal guidance, feedback and assessment systems, whereas assignment sheets must be prepared separately and are not integrated with the simulation systems. Based on our work so far, we propose the flexible assessment system Antares (adaptive network-oriented tracking-based assessment for real-time educational simulations), which builds on a service-oriented architecture that targets real-time simulations. It allows for a flexible separation and reusability of several components of the learning system, such as the immersive 3D environment itself and the assessment engine. While a generalized, annotated immersive 3D environment can be reused for several similar experiments, the external assessment engine does not only offer enhanced, adaptive assessment measures—based on user actions and object states—but lets the teacher provide custom assignment sheets, which can be injected to compatible display elements within the immersive environment. First applications on a pendulum experiment demonstrate that the

J. Maderer (✉) · C. Gütl
Graz University of Technology, Graz, Austria
e-mail: joachim.maderer@student.tugraz.at

C. Gütl
e-mail: c.guetl@tugraz.at

C. Gütl
Curtin University, Perth, Australia

R. Babo et al. (eds.), *Workgroups eAssessment: Planning, Implementing and Analysing Frameworks*, Intelligent Systems Reference Library 199,
https://doi.org/10.1007/978-981-15-9908-8_7

approach can be integrated with low expenditure, and the evaluation algorithm is able to match complex real-time measurement procedures.

Keywords STEM education · Assessment · Guidance · Behavior · Physics · Immersive environments

7.1 Introduction

Worldwide, politicians, economists and other stakeholders are still stressing the need for better STEM education and increased numbers of graduates in those fields. While the emerging field of desktop and mobile virtual reality (VR) technology offers new chances for highly immersive, personalized learning environments, proper assessment and guidance concepts must not be neglected and should be grounded in adequate theoretical and practical considerations.

Basically, most relevant learning theories can be summarized into three broad categories, including behaviorism, cognitivism and social as well as individual constructivism [1]. While behaviorism is only comparing desired to actual outcomes and reinforcing them with rewards (or penalties), cognitivism is deeply concerned with the internal mechanics of the human brain; constructivism advocates the idea that every individual, respectively a social group, must construct its own knowledge [1]. Especially, the latter has caught the interest of STEM educators over the last years, stressing the assumption that students may learn best with inquiry-based learning methods, thus formulating research questions and conducting appropriate experiments on their own. Based on that, they can then create their own knowledge (whether it be in socio-cultural groups or as an individual) [2]. Unfortunately, little research and literature are available revealing significant results that this approach actually works and there are several proponents against such constructivist ideas [3]. Particularly, the need and expectations of new generation of learners regarding feedback and guidance is a critical aspect. More open, competence-oriented approaches might work for advanced learners (*cf.* [3]). But in addition to that, authors also express the importance of clear guidance and timely feedback [4, 5].

However, this kind of flexibility is often impossible to achieve with inhomogeneous student groups, which teachers are usually confronted with on all levels of education. While implementing real experiments for a large group of students comes with several problems (available budget, dangerous settings, etc.), virtual experiments and simulations benefit from recent advances in immersive technology (*cf.* [6, 7]). Not only have simulated environments be found to increase conceptual understanding, but they can also improve experimental skills and add additional advantages such as the visualization of abstract concepts (e.g., forces or electromagnetic fields); yet, instrumenting them pedagogically sound is a challenge on its own [6, 7].

Physics teachers appear to be well aware of the existence of physics simulations and use them primarily to explain physics concepts to students [8]. However, a closer look at the most popular tools—which are known among physics teachers—reveals that these tools hardly implement any automated assessment concepts and by no means adaptive guidance and feedback procedures. Besides, the tools are usually only 2D and thus low immersive (*cf.* [8]). Hence, if teachers want their students to experiment with such tools during self-regulated learning phases, they either must use—if available—the included written materials or create their own traditional assignment sheets. This approach does not only increase the chances of distraction and lack of pedagogical control but more importantly misses the vast potential for smart solutions to deliver feedback immediately and tailored to the individual needs and pre-knowledge of students (*cf.* [5]). The latter can not only increase the efficiency of learning, but it also reduces the stress teachers are facing inside the classroom during the active review of student exercises, allowing them to focus on students who need further help.

Another observation worth mentioning is that a lot of teachers—not in STEM education particularly—lately use popular online services for assessment which do not comply with established exchange formats such as IMS QTI.[1] This does not only hinder the development of open educational resources; it further introduces critical decisions on where to create questions or invest the institutions budgets or even personal money.

The situation described so far emphasizes the need for a flexible and adaptive assessment system that can be integrated with different learning environments, is able to recognize skills (procedures), as well as the application of knowledge, and is centered around reusable learning content and assessment items. With the increasing potential of using immersive educational environments—such as Maroon [9]—to motivate students, our workgroup has already been occupied for several years to not only integrate behavior-based assessment into such environments but to specifically design an interface that is highly flexible in terms of reusability and support for different (immersive) learning environments. As a result, we consolidate our already invested efforts and introduce the Antares framework as part of this paper. It stands for adaptive network-oriented tracking-based assessment for real-time educational simulations.

The aim of this chapter is to formulate updated generic requirements for a flexible assessment system to be used with immersive environments, describe the Antares framework, as well as its latest conceptual changes, and showcase a working prototype that features a virtual, highly immersive physics experiment, implemented in Maroon and modeled in accordance with the framework.

The remaining chapter is structured as follows: The next section summarizes related work in the context of assessment and guidance in virtual learning environments. Section 7.3 introduces the Antares framework, its overall technical approach and details related to the evaluation process. Section 7.4 presents selected showcases of assessment rules and feedback messages in the context of a VR pendulum experiment. Section 7.5 gives an overview of future developments in this research area, whereas Sect. 7.6 discusses future work and concludes the chapter.

[1] https://www.imsglobal.org/question/index.html.

7.2 Related Work

The range of approaches to assessment and interactive guidance is broad and can be examined from different perspectives. It includes features like adaptivity to learner performance and preferences, flexibility in terms of content and environment and the type of assessment used, such as fixed-response questions, natural language processing or tracking of user behavior. Beyond that, the question of what should be assessed is of interest too, whether it be related to knowledge, skills or competences.

More simple approaches to formative assessment are based on the usage of traditional e-assessment items, such as quizzes. Early examples in the context of immersive environments, such as 3D virtual world platforms, consider it as extended viewer for established *learning management systems (LMS)* [10]. Tools such as Sloodle aim toward flexible solutions to integrate existing content from Moodle into the immersive environment Second Life and return accomplishments and assessment data to the LMS; QuizHUD is another option to create simple interactive e-assessment items in-world [11]. However, more sophisticated scenarios have been implemented as well, with the drawback that programming skills among all relevant technological platforms are required [11]. There are also approaches that do not consider the immersive environment as mere viewer but rather as the central learning setting that is supported by external systems [10].

In the context of learning games, approaches to 'micro-adaptivity' that support the learner in finding the right solution through contextualized pedagogical agents [12], as well as hidden evidence mechanisms for the development of competencies, have been studied [13]. The authors in [13] describe a 'stealth assessment' concept that relies on Bayesian networks to calculate probabilities that certain competencies have been reached. Accomplished goals inside the game are used as evidence and are mapped to the lowest level of the hierarchical competence model. This happens under the assumption that competencies cannot be measured directly, using an approach referred to as evidence-centered design (ECD), originally proposed by Mislevy et al. (as cited in [13]). It has been tested in several game-based settings.

A related concept named *competence-based knowledge space theory (CbKST)* has been applied in the design of the learning game 'Elektra' as described in [12]. While the game itself is driven by a catching story, several physics experiments must be solved in order to progress inside the game. The learner's competence status is represented as an ontology, where the demonstration of specific skills and knowledge are considered as necessary or sufficient prerequisites to other nodes in the model. Based on these relations, competence states can be assumed, and the game can be slightly adapted in a non-invasive manner. The latter is called 'micro-adaptivity' and occurs for example by instrumenting a pedagogical agent that gives hints to the player and acts as a hidden teacher [12].

Both concepts, however, focus on the overall psychological and pedagogical research concept and appear as rather hard-wired, inflexible solutions. Especially, the achievement of tasks is described as specifically implemented inside the game engine. There is no mentioning of a generalized way to detect the desired behavior of the player.

When it comes to strong guidance, *intelligent tutoring systems (ITSs)* have been in the focus of researches and developers for several decades. Following the authors in [14], an ITS can be summarized as a computer system that provides adaptive learning content, questions and answers, as well as feedback and hints by modeling a student's psychological state on multiple dimensions. Typical architectures consist of four types of models, including the *domain model* that is responsible to encode the expert knowledge of a subject area and provide solution paths; the *student model* that is used to record individual choices, misconceptions or knowledge; the *tutor model* that makes decisions regarding the next steps and selects the proper learning content; and the *interface model* that communicates with the user [14, 15].

According to the authors in [14] and [15] there are several types of ITS that can be distinguished, particularly in terms of the logic used for the domain model: *Model-tracing tutors* offer step-by-step evaluation of the learner's interactions with the system. This is achieved by a set of production rules that form a solution graph. Each rule represents a possible intermediate step, whereas possible transitions can be categorized as *valid* (directly leading to a new state), *buggy* (recoverable, but not the best approach) or *erroneous* (unable to proceed with this operation). The transitions could be used to provide immediate feedback to the learner. Another intelligent tutoring concept is *constraint-based tutors* which are only focusing on the correct result, expressed as conditional constraints, rather than an exact solution path. But it is still possible to give useful hints based on violated constraints [14, 15]. Such systems can be easier built and executed than model-tracing tutors [16]. However, more recent *example tracing tutors* can offer functionality similar to model-tracing tutors with less effort and do not require programming skills. Instead, experts provide possible solutions which can be automatically compared to the learner's solution path [17]. In addition, there are also dialogue-based tutors which are leading the user to the correct solution by providing a conversation with an intelligent agent as well as different concepts with decision models based on Bayesian networks [14, 15]. Other approaches based on data mining and machine learning are investigated as well, e.g., in the context of inquiry-based science education [18, 19]. However, most of such systems work as rather simple enclosed web or desktop applications, in contrast to complex, open-ended immersive environments.

A more recent approach that targets immersive 3D environments and draws ideas from example tracing tutors is the *Semantic-enabled Assessment in Virtual Environment (SAVE) framework* as described in [20, 21]. It works with generalized traces of user actions, which are first demonstrated by an expert and represented as graphs. These action traces can further be annotated to allow deviations, such as optional actions or the arbitrary order of a partial set of actions. As the system relies on a semantic description of the corresponding virtual environment, it is possible to automatically generate meaningful assessment messages based on an enhanced graph matching algorithm. But it is not possible to provide custom feedback messages. However, the entire design is focused on the assessment and training of highly procedural skills. The system is not concerned with changes in the environment, the state of other objects or past achievements, that might influence the current assessment process, although the authors discuss these shortcomings [20, 21].

Finally, the author in [22] has discussed behavior in virtual learning environments form a theoretical perspective. He proposes a taxonomy called *BEHAVE* which categorizes user actions into a hierarchical model of *goal*, *constitutional* and *functional acts*. Functional acts represent the most basic actions a user can perform inside the virtual environment, which could be classified into *Gestural*, *Responsive*, *Decisional*, *Operative*, *Constructional* and *Locomotive* actions. Constitutional acts combine one or more functional acts to logical activities, and goal acts could be considered as desired overall outcomes. Based on this theoretical foundation, the *Action-based Learning Assessment Method (ALAM) framework* is introduced as a method to define parameterized action sequences and their required order by applying relationships known from the field of project management (start-finish, etc.). This approach shows many similarities with the SAVE framework (*cf.* [20, 21]), is also focusing on procedural skills and is not concerned with the state of surrounding objects.

To our best knowledge, literature does hardly reveal further relevant frameworks for student assessment in exploratory learning environments, let alone in such specific contexts as immersive environments. Following the authors in [18] and [23], several approaches have been studied within the last decade which focus on the assessment of skills and competencies, but further research and improvement are required, particularly in terms of practical application. However, most approaches rely on probabilistic decision trees such as ECD or other forms of numeric metrics, whereas most of the underlying activity tracking appears hard-wired. The authors in [23] further describe an assessment concept for immersive environments that investigates correlations between the skills assessment of an expert and quantitative metrics which are collected from more general in-world activities.

Focusing on our primary domain of interest, namely STEM education and physics in particular, theoretical concepts related to inquiry-based learning, more precisely experimental skills, are better related to our research. The authors in [24] have found evidence that the assessment of experimental skills relying on process-based observations could in certain dimensions be superior to the assessment of results alone (e.g., measurements on an answer sheet). Based on that, we believe that action-based models should be preferred, in contrast to the arguments expressed by proponents of constraint-based tutoring models (*cf*. [16]).

7.3 The Antares Framework

The literature review reveals that simple assessment approaches in immersive environments, such as fixed-response questions, have already been integrated with external systems. The observation and analysis of behavior—in order to determine competence levels and provide adaptive guidance to learners—have been the focus of pedagogical and psychological studies. However, the described prototypes appear to rely on hard-wired in-game achievements as there is no mention on a generalized and flexible approach to detect player behavior in different situations.

Constraint-based tutors seem less applicable to simulations of virtual laboratories as they are not concerned with the procedure itself but only with the outcomes. Rule-based and model-tracing tutors are more promising in that sense, although there are issues related to the support of open-ended environments and rather limited task domains (cf. previous section). The few examples that focus on behavior-based assessment are only focusing on sequences of user actions and generic feedback and do not consider the state of the surrounding objects for an overall behavioral judgment. Based on that, we conclude that a hybrid approach featuring aspects of model-tracing tutors as well as constraint-based tutors seems best suited for open-ended inquiry-based learning, where exact (sometimes interchangeable) experimental procedures constitute a key element.

In order to address the aforementioned issues, we propose the Antares framework as a platform-independent, flexible assessment approach that is focusing on behavior-based assessment in different situational contexts, including traditional assessment and content items among various feedback types and interconnecting with external systems, e.g., competence models, to provide adaptive guidance to learners.

Regarding the development methodology, we are following an iterative research cycle that corresponds with the methods of design science research (*cf.* [25]). We are currently in the stage of a major development cycle and are preparing didactical studies and user questionnaires for a large-scale university physics course in order to evaluate our approach. Beyond that, adapted assignments will further be tested in secondary school education.

The following sections will introduce requirements and design decisions on a generic level, explain the architecture of the Antares ecosystem and give details on the actual assessment logic.

7.3.1 Requirements and Design

Based on the extensive literature review and our early contributions [18, 26, 27], the most important requirements for a flexible assessment system can be summarized as follows:

- *Support for different platforms*: The need for this requirement is manifold. On the one hand, it has been outlined that domain experts are challenged with different technical approaches and would therefore benefit from a unified interface. On the other hand, if certain learning environments cease to exist, a decoupled approach helps to preserve the invested work and knowledge. Besides, the same assessment logic can be used for complementary systems (such as desktop, mobile, virtual reality or augmented reality versions).

- *Independence of actual learning content*: Since several recurring principles can be found in different assignments and environmental constellations, the entire assessment plan or at least significant parts should be reusable for similar simulations that are built from the same repository of items or are semantically compatible.
- *Flexible assignments*: Simulations and equipment might be reused for different assignments that are not comprehensively known in advance. Instructional designers should therefore be enabled to inject any desired learning content (supportive materials, instructions) and formulate adequate assessment rules. This may also include the manipulation of the environment to set up certain training conditions.
- *Feedback and scaffolding*: Learners should receive well-timed feedback based on their actions in different situations (from immediate feedback, over intermediate summaries to final debriefings), including scaffolding techniques to guide the learners toward their best possible achievement.
- *Adaptivity*: Students with different prerequisites should be able to successfully complete the learning activity without getting frustrated. Therefore, appropriate feedback measurements should be provided under consideration of the current competence state and personal preferences of the student.
- *Behavior-based assessment*: The system should be able to assess learners based on their actions in conjunction with the situational context that is defined by the surrounding environment, thus including the current state as well as state changes of nearby objects and the impact those information has on the learners in terms of observation, reaction and reflection.
- *Collaborative learning*: A group of learners should either be able to complete tasks together or separate the workload with regard to the given assignments. The assessment system should be able to evaluate subject-related achievements as well as aspects related to the social dynamics of collaborative activities.
- *Reporting and learning analytics*: Teachers and instructors should be able to receive live reports as well as final reports about the progress of either individual students, groups of students as well as system-wide assignments.
- *Interoperability with external systems*: Supportive services such as grade books or competence models, but also additional learning materials, may be stored in external systems (such as LMS). It should be able to interconnect these services with the assessment system.

7.3.2 Architecture

In comparison with traditional e-learning objects, Antares places an additional layer between the immersive environment as the actual learning object (LO) and the APIs and resources usually provided by learning management systems. This decision is grounded in the fact that simulations represent complex items which can hardly be maintained by domain experts themselves. Externalizing the assessment process

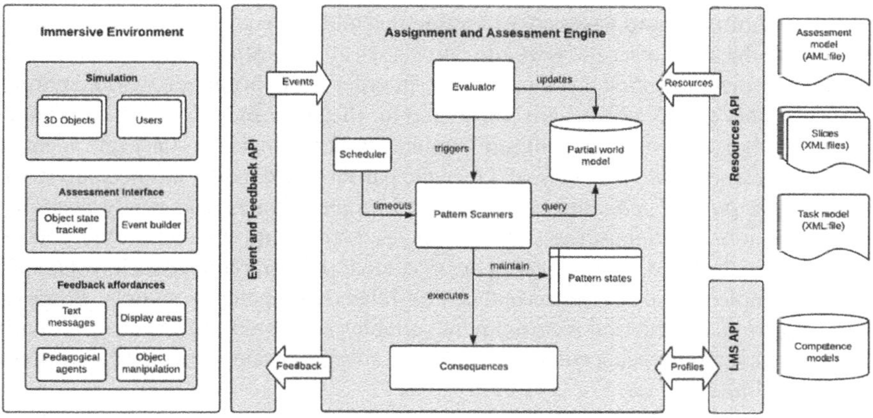

Fig. 7.1 Technical architecture of the Antares framework

increases the flexibility by enabling instructional experts and teachers to adapt assessment rules or entire assignments without changing the content of the simulation itself.

Based on early drafts and conceptual architectures that have been reported before [8, 26, 27], a simplified technical architecture (Fig. 7.1) has been created, summarizing the most important terms and concepts:

- **Immersive environment** refers to the technological platform that is used to create the simulation. Being the result of a joint process between technical and domain experts, the *simulation* should not directly address didactical and pedagogical concerns. However, a generalized event tracking mechanism (*assessment interface*) is supposed to track the state of all relevant (nearby) objects, as well as actions performed by users. Both, state changes and user actions, are compiled into events (*event builder*) which are handed over to *the event and feedback API*. In order to process incoming feedback, several *feedback affordances* are considered:
  - text messages are the most basic form of feedback and should be displayed in different colors based on their meaning (e.g., success or mistake);
  - display areas are special in-world objects that can display dynamic content ('slides') which are based on a simple markup language;
  - object manipulation refers to the ability to change certain object properties or trigger special actions from the context of the assessment system. Possible use cases include step-by-step demonstration of procedures as well as prepared setups for certain assignments (placing several entities inside a virtual laboratory);
  - pedagogical agents might be instrumented to deliver the feedback to increase motivation.

- The **assignment and assessment engine** resembles the central reusable components of the approach. The *evaluator* processes all incoming events and updates a *partial world model* on a per-user basis. In order to support open world environments, only nearby objects are considered for the assessment process. The aim is to provide a context-dependent assessment experience that takes only into account what a learner could be aware of. Once the partial world model has been updated, all active *pattern scanners* are triggered to match against the newly available information. If certain series of state changes, actions and correlated object states have been detected, *consequences* are executed, resulting in feedback responses, task completions or evidence updates on related competence models. Progress information and internal measurement variables are stored in *pattern states* until a pattern scan has succeeded or failed. The *scheduler* supports the scan process by handling *timeouts*.
- The **Resources API** provides access to important working files, including:
  - the *assessment model* which defines patterns and consequences ('assessment rules') to be used in the current learning session;
  - different XML-based templates for *slides* in order to inject task descriptions and learning content (such as text, images and forms) into the immersive environment;
  - and a *task model* that contains a hierarchical list of tasks and subtasks. It is used to describe an explicit (visualizable) or implicit (hidden) assignment structure to group and activate assessment patterns relevant in the current stage of the learning experience.
- Finally, via the **LMS API**, information regarding the current user profile is exchanged. This includes, but is not limited to individual competence models, which are used to adapt the feedback suitable to the current learner and to store evidence associated with the increase of competences.

Further details regarding the communication protocol between immersive environment and assessment system, as well as the internal mechanisms of the assignment and assessment engine, will be explained in the subsequent sections. In terms of collaborative learning, the fundaments of Antares have been designed for multi-user environments from the very beginning. Basically, all members of a work-group can contribute to the solvation of a problem, and the assessment system will be able to recognize who has performed which activity or action sequence. The different perception perspectives of the different users are considered by the communication protocol and stored as unique layer within the partial world model. However, at our current development stage, we are focusing on single-user assessment in large courses to evaluate effectiveness of the assessment and feedback mechanisms. The following iterations will then further investigate the applicability in open collaborative environments.

7.3.3 Assessment Interface

The data exchange between immersive environment and assessment system is conceived as an asynchronous, bidirectional communication channel. Supported message types include *event* messages, which are built by the immersive environment and forwarded to the assessment system, as well as a range of *feedback* messages. The latter must be interpreted by appropriate feedback plugins which need to be implemented once per platform. Earlier prototypes [26, 27] have already shown that adequate real-time performance is achievable by connecting the assessment system through a web service layer. However, due to the latest additions regarding timeout-based behavior detection and practical reasons, the previous approach has been dropped in favor of a modernized and simplified protocol that will make use of JSON and WebSocket technology.

Amount of Data

A critical aspect is the proper selection of the sufficient data that should be transmitted to the assessment system. Since the requirements include the support of open-ended large-scale environments and simulations that run in real time, sending the entire state of the environment multiple times per second is not an option. For that reason, a strategy was developed that only includes in-world objects which are in perceptional range of the assessed users.

In addition, the object states (object properties) have been categorized into three nested levels of importance:

- *Dynamic* changes occur within milliseconds and are not supposed to be transmitted continuously. While simulations will often depend on such values, their usefulness was found neglectable for the assessment process—without any other (slower) changes or users actions happening at the same time. A good example for a dynamic change is the current deflection of a pendulum, without pressing the button of a stopwatch at the same time. However, zeros of derivatives of the deflection, such as velocity or acceleration, could constitute 'slow changes' which could be of interest together with all other values.
- *Changeable* data is considered to represent customizable settings or state that is inherent to the system. Such values may change because of direct user actions or as a result of random or special events within the simulated environment (cf. previous point). They are not expected to change too often but are relevant for behavioral assessment.
- *Full* data updates include the entire state and description of an in-world object. Besides *dynamic* and *changeable* data, this level contains static information, such as the type of the object or other observable parameters (e.g., color, size, etc.). This level of data is only transmitted whenever the objects come into the perceptional range of the user.

7.3.4 Assignment and Assessment Engine

The Antares reasoning engine could be described as an enhanced rule-based system that borrows aspects from classical expert systems [28], regular expressions for string pattern matching [29], as well as first-order logical languages such as data querying languages [30]. It is a significantly refactored and extended fork of the *SOFIA Evaluation Engine for Game-based Learning* [26], an early prototype developed by our workgroup.

The evaluation engine is based on graph structures, which are heavily inspired by regular expression algorithms. That means assessment rules are formulated by defining search patterns for specific or generalized sequences of actions as well as environmental changes. These patterns are further extended by defining an arbitrary number of consequences such as feedback responses or internal state and measurement features (e.g., elapsed time between actions). Therefore, the remaining work will refer to these assessment rules as *behavior patterns*.

Design Considerations

In comparison with other intelligent tutoring systems, as described in the related work section, it was a design decision from the very beginning to not formulate strict solution graphs in terms of finite state machines but rather generalized behavior patterns. Under these circumstances, we consider it possible to reach a level of flexibility that is not limited to a mere replacement of graphical representations, but the assessment can be used in different open-ended situations that share a common terminology. To give a more concrete example, consider a practical driving exam: If one has understood how to drive safely based on the current environmental situation, he or she can drive anywhere. The same is true for the assessment capabilities of a driving instructor or examiner. Thus, if it is possible to replace the entire situation it should not be necessary to edit all behavior patterns, they should be valid for almost the entire problem domain, or at least several parts should be easily reusable.

While the implications stated above could partially be solved by a typical rule-based decision system (conditions lead to consequence), it still misses out the exact transitions that lead to the desired state. To stress the driving example again, the sequence of actions a driver performs immediately before changing to another lane is of uttermost importance. But hardly can it be assumed that not looking into the mirror is a mistake if there is no intention of changing the lane. This leads to the introduction of search patterns that match against incoming user actions and environment changes. Like the formalism that is associated with regular expression, quantifiers for repetitions, subgroups and back references to previously matched fragments and patterns are used to determine occurrences and quality of activities.

Finally, the behavior of independent, reusable objects that exist in close range to the user might also be relevant in relation to the actions performed. While there is not necessarily a direct (technical) relation between them (e.g., stopwatch and pendulum), the assessment system needs to establish a correlation between both objects and the user's actions performed on them. Other approaches in the context of procedural skills have only mentioned the idea of explicit object states as additional parameters to actions or environment conditions that must be met at a certain point of the action sequence. Our approach reaches beyond that point in two ways: First, we consider changes of object states as first-class events equitable to user actions (timeouts alike). Second, our partial world model can be queried at arbitrary points of the match process, allowing the binding of objects which are related to the current behavior into the current scan process. Here, the most likely intuition will be that the closest, most recently used object of a certain class is that one which is currently observed and considered by the user, although looking directions could provide additional evidence.

Functional Principal

A *behavior pattern* consists of a linked list of abstract nodes, which are referred to as fragments. Each *fragment* is responsible to evaluate a certain aspect against the current environmental context (partial world model and pattern state) and can either *succeed*, *fail* or report the requirement of *suspension*. The latter means that the evaluation process cannot be finished at this time, which will usually result in a suspension of the entire behavior pattern.

The evaluation engine itself creates a *pattern scanner* for each behavior pattern and can manage multiple 'recognition contexts' for each scanner. Since a behavior pattern can begin with any incoming event, multiple instances of the same scan process might be active at the same time.

Assessment Modeling Language (AML)

Antares has been extended with its own language to enable rapid model development. While a graphical assessment editor is currently being developed, the language is supposed to evolve faster and will probably constitute the better choice for computer professionals and ambitioned domain experts. The overall structure of the language as well as its procedural elements have been inspired by the Python[2] programming language, whereas the data querying clauses have been borrowed from structured query language (SQL). The approach to event sequencing is influenced by regular expressions as used in string pattern matching. Each node of the pattern structure (as described in the previous subsection) is reflected by different keywords and expressions used in the language, which will be introduced and discussed in the following subsections. Concrete examples of source code will be given in the next section together with working showcases.

Primary Matches

Based on literature comparison and previous experiments, three types of events (see Table 7.1) have been categorized as important 'primary actions' which are supposed to be awaited until they occur. It means that pattern scans in progress will be kept in a waiting state ('suspended') until the relevant information becomes available. This enables the engine to not only analyze proactive behavior, but also reactive and

---

[2]https://www.python.org/.

Table 7.1 Primary match expressions, waiting for incoming events

Type	Language construct	Explanation
User action	Match action	Movements (and related actions) as well as object interactions are considered explicit user actions. Actions can contain additional information, such as which objects where involved in the operation
Object state changes	Match property changed	The state of objects can change based on random or system-driven events within the simulation or as a side effect on dedicated user actions
Timeouts	Match timeout	Observing that someone is not working toward a goal can be considered an 'action' by itself. Thus, patterns block on timeouts and determine if task goals have not been reached at that point. Suspensions can be defined to prevent timeouts from occurring during promising attempts toward a specific behavior pattern

reluctant behavior, such as reactions to possible observations (environment changes) as well as feedback in the case of disorientation.

Consequences

In the context of Antares, consequences are understood as the entirety of commands that cause either internal or external changes to the environment. This includes feedback messages, which are sent to the immersive environment, as well as collected assessment data, but also the update of internal variables that support the match process of a specific pattern scanner instance. As specific patterns could start with any incoming event, several concurrent instances of the same pattern must be managed at the same time. Reading or writing internal variables will therefore access the local data context associated with that instance of the pattern scanner that is currently evaluated. Blocks of consequences are either declared explicitly as a new node or implicitly by using at least one of the procedural statements as summarized in Table 7.2.

Groups and Sub-Pattern Matches

In order to allow more complex behavior sequences to be detected, sequences of primary action matches and consequences can be grouped together and annotated with quantifiers (see Table 7.3). Thus, it is possible to match repetitions of the same sequence of events until the subsequent node evaluates to true. Another option is the matching of entire patterns that are defined elsewhere. As all successful matches will be recorded within the partial world model, simple process evaluations can be reused and combined to more complex behavior patterns.

Querying Objects and Previously Matched Patterns.

Table 7.2 Overview of procedural commands available in Antares

Statement	Explanation
Feedback	Sends a textual feedback message to the immersive environment. The type of feedback can be distinguished between 'success', 'hint', 'warning', or 'mistake', which should result in different representations within the immersive environment (e.g., color scheme)
Display	This command is used to transmit a certain content item (e.g., an XML-based slide) to an in-world target (such as a presentation wall). The target would usually be selected by object query operations
Set	Creates new custom variables or updates existing variables or object properties. Existing higher-level variables have precedence over local variables
If, elif, else	Branch statements which work the same way one is used from other procedural languages (including logical conjunctions with 'and' and 'or')
Record	Stores one or more output variables that will be available once the current pattern instance succeeds and is stored in the partial world model. This storage mechanism is used to describe the quality of the recognized behavior pattern. It can be retrieved by other pattern scanners if past behavior needs to be considered for the current assessment process
Enable, disable	These commands are used to activate or deactivate one or more pattern scanners. Once a pattern scanner is deactivated all pending search paths (pattern instances) are destroyed
Suspend, resume	In comparison with the 'disable' command, 'suspend' only freezes a pattern scanner, including all active search paths. Pending timeouts will not fire during suspension but will immediately be evaluated if already due on resumption. Suspensions and resumptions are counted if used multiple times and failed patterns will automatically decrease the counter of associated suspensions

Table 7.3 Group-based match expressions and sub-patterns that allow the formulation compound behavior patterns

Language construct	Explanation
Match group	All nested fragments must match successfully in the order they have been declared. If a subordinate fragment suspends its evaluation due to unavailable data, the group node will be suspended as well. Optional quantifiers may determine the minimal and maximal allowed repetitions
Match first	The nested fragments represent alternative branches that will be evaluated in declaration order. The first successful subordinate fragment completes the evaluation process of the container
Match any	Like the previous group type, except that all nested nodes are evaluated instantly if a predecessor would suspend the evaluation process
Match pattern	Awaits the successful detection of another named pattern. The timestamp of the recorded pattern cannot be older than the current timestamp. Results may be stored in local variables (similar to query operations, see next subsection)

Since the entire state of objects, currently known to the user, is stored in the partial world model, objects relevant to the assessment process as well as previously detected patterns can be queried based on conditional clauses and in preferred order. The results of both types of queries can be limited and stored in local variables. It is either possible to access property values of those objects as needed or to use the query results as input for other match expressions. For instance, it is possible to limit the range of allowed objects a user action or property change event is matched against.

As query expressions represent another fragment type like 'primary actions', the default behavior requires at least one result at the time the fragment is evaluated. Otherwise, the current subtree of the pattern will fail. However, an optional annotation is considered at this time to allow the progression to the subsequent fragment if the query yields an empty result. This could be desired in some situation, especially if the availability of certain objects is optional.

Furthermore, query expressions can be used quite flexible as syntax allows the placement of this expression in two ways:

- *Ahead of primary actions*: This is useful if primary action matches depend on the existence of certain objects. Since queries can be considered immutable operations and the partial word-model could be subject to changes while the engine waits for relevant incoming events, it is considered to automatically reevaluate preceding queries and conditions up to the last mutable operation;
- *Nested*: This case is relevant if the binding of certain objects must correlate with key actions or changes to the environment. Whenever a primary action fragment features nested fragments, such as the query expression, it can only evaluate successfully if all subordinate fragments succeed as well. The entire evaluation step is then performed atomically and the found object states as well as the matched actions share the same timestamp.

The described querying features are currently implemented rather rudimentarily. At the time of writing this chapter, it has already been planned to enhance the partial world model to support rich semantics. This will be achieved either by relying on existing standards and libraries, such as Open Web Ontologies (OWL),[3] or by implementing a custom approach, if necessary. The expected main advantages include the description of object types and associated actions in generic terms, which will enable the definition of type hierarchies for objects and actions. That means it will be possible to reason about certain characteristics indirectly or match against entire classes of actions and properties instead of specific actions.

However, the static nature of this kind of reasoning can only infer about facts that have been declared and are available at the exact time of reasoning. Since our approach is also concerned with the sequence of transitions that leads to a new state of the partial world model, we intend to infer the type of an object not only by what features it provides but also on how it behaves. For instance, a simple pendulum, which is only a special case of an oscillating system, can also be identified by looking

[3]https://www.w3.org/OWL/.

Table 7.4 Conceptualized annotations applicable to pattern fragments that add metadata on how to process the nodes (i.e., constraints and side effects)

Annotation	Explanation
Snapshot	Captures the state of the partial world model at the time the node is successfully evaluated. Subsequent access to variables can be prefixed with the snapshot name to obtain the intended object status
Task	If applied to the root element of the behavior pattern, the pattern will only be activated and/or available if certain tasks of the future task model are active. Nested nodes can be annotated as well, allowing to ignore specific parts for some tasks, e.g., introductory help messages
Optional	Optional nodes will not break the match process if they fail
Competence	Feedback messages and/or certain parts of the scan process will only be evaluated if the competence levels of the user match the desired levels of the annotation
Limit	Feedback messages and/or certain parts of the scan process will only be evaluated if the number of reruns matches the given limits. Thus, different help messages or feedback types can be used if problems continue to exist

at its periodic changes of location (respectively rotation), velocity and acceleration. As this is also true for the spring pendulum, a generalized assessment model could work for both scenarios.

Annotations

Finally, the entire definition of the behavior pattern as well as almost all types of available nodes can be annotated with different modifiers that introduce side effects, determine the applicability of the node in certain situations or influence its execution behavior. All annotations that have been conceptualized so far are explained in Table 7.4, although not all annotations have yet been implemented.

7.4 Slide Templates

For the realization of exchangeable in-world content, a simple XML-based template format has been designed to describe and visualize basic textual content (headings and paragraphs) as well as images and simple table structures. Although some generic layout parameters can be provided, the concrete representation is left to the target platform, thus ensuring responsive design and native visualization. In addition, common input fields like textboxes, checkboxes, radio buttons and selection lists are provided to enable the author to prepare answer sheets or quizzes. Buttons can be defined with custom action names. Once a slide has been loaded into a compatible in-world object, the slide itself is supposed to behave like any other object that is part of the environment. Hence, the usage of buttons will cause the creation of new events, containing the defined action name as executed user action and all current values of all form elements as updated object state. However, for future updates, we intend to integrate an interface to existing e-learning items based on standardized formats such as IMS QTI. These items can then be automatically rendered as internal slides and sent over to the immersive environment.

The custom XML format was conceived as a lightweight format to be easily converted to a proper representation on the target platform. But we are also considering the usage of native HTML 5 documents as alternative content format for future releases. This would improve the interoperability with existing learning content and enable the dynamic placement of background knowledge and instructions inside the immersive environment. However, the behavior of forms would require special handling, since submissions and changes in form fields have to be converted into events that are compatible with the Antares communication protocol.

Despite the format actually being used, slides should always be considered templates that are rendered to concrete slide instances, once the 'display' command is executed. This enables placeholders to be filled with meaningful data, including, but not limited to, usernames or custom parameters in dependence of random assignment deviations, etc.

7.5 Proof of Concept

As our research is focusing on explorative learning settings in STEM education, and the assessment system has been designed with the support of highly dynamic interactions in mind, we have chosen a simple pendulum simulation as long-term study object to proof the technical applicability of the approach.

The simulation of the simple pendulum has recently been reimplemented and improved with the Unity[4] game engine on top of Maroon framework [9]. In addition, a collection of scripts and components has been developed to simply connect simulations in Unity with the Antares assessment engine. Besides text-based feedback messages, a presentation area for rendering incoming interactive content (slide objects) has been added. It makes use of the built-in user interface components of Unity.

7.5.1 Learning Situation

The assumption is that learners have just been introduced to the basic concepts of oscillations and are requested to investigate the behavior of a simple pendulum. Following an inquiry-based research cycle, it would first be necessary to state a proper hypothesis. That means it is necessary to describe the expected behavior and explain the theoretical considerations involved. While the assessment framework could provide injected in-world questions for this kind of assessment, it is not the

[4]https://unity.com/.

intended focus of this research, nor is it necessary to include all aspects in all assignments at the same time. Consequently, we have chosen the following aspects to be considered for the guidance and assessment process (*cf.* [8]):

- *Conceptual understanding*: At the beginning of the exercise, the learner is asked to determine the current frequency of the pendulum by measuring the cycle duration. At this point, conceptual errors regarding the concept of a cycle duration can be intercepted, as can be wrong formulas. Advanced students can also improve the measurement quality by counting multiple cycle durations.
- *Systematic measurements*: Once it is ensured that the student can correctly determine the cycle duration and frequency based on the current settings, it is necessary to investigate the dependent variables by systematically varying only one parameter at a time. Based on this information, the assessment system can infer what situations have been systematically analyzed.
- *Observations*: If the learner has correctly observed the behavior based on proper systematic measurements, it should be possible to ask reflective questions about the behavior of the system.

7.5.2 Starting the Simulation

Once the student enters the location of the experiment, a welcome message is displayed, and the appropriate slide is transmitted to the simulation. The latter consists of written instructions, an input field for the calculated frequency and a submit button (see Fig. 7.2).

The corresponding behavior patterns are shown in Fig. 7.3. First, a slide is loaded from an XML file into the session scope. Second, the pattern 'init' is defined to

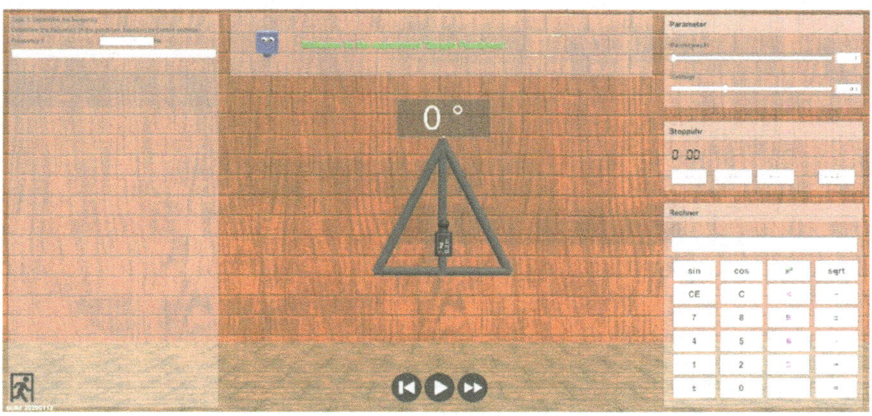

Fig. 7.2 Simple pendulum simulation, including parameters, stopwatch, calculator and dynamically injected instructions on the left-hand side

```
load slide "pendulum_task_1.xml" into slide_1

pattern init:
    match action enter
    query objects \
        where class == [Display] \
        into board
    display slide_1 on board
    feedback success "Welcome to the experiment 'Simple Pendulum'."
```

Fig. 7.3 Behavior pattern ('assessment rule') used to set up the assignment

describe the necessary consequences that follow the user action of entering the scene. Immediately afterward, a 'query objects' expression is used to search the partial world model for an object that can display slide content. The result of this query is used in conjunction with the slide template to send a concrete slide to the simulation. Finally, the welcome message is sent.

7.5.3 Measurement of the Cycle Duration

In order to determine the frequency of the pendulum, four independent objects must be considered, namely the pendulum itself, the stopwatch, the calculator and the form to submit the result. The learner is supposed to pull and release the pendulum and press the start and stop button of the stopwatch at the same position. However, multiple cycles are allowed and will be considered good measurement practice by the assessment system. The system will also warn the student if measurements exceed a certain tolerance range. The screenshot in Fig. 7.4 shows a feedback message that informs the player about a conceptual mistake.

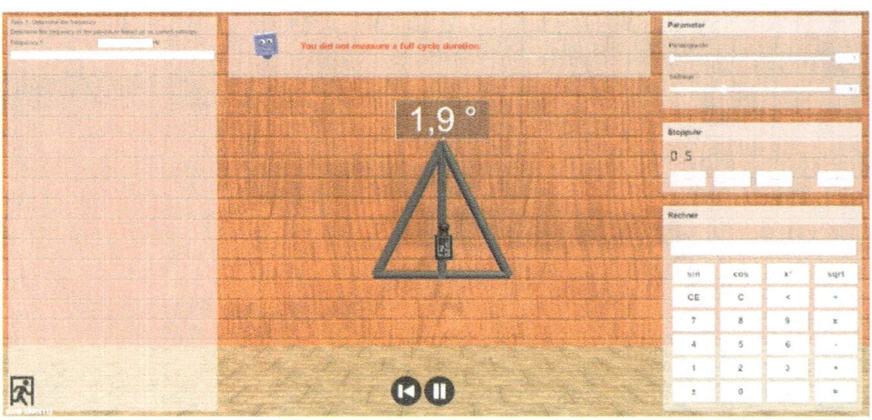

Fig. 7.4 The student is informed that the measurement was conceptually wrong

```
14 pattern Measurement:
15
16     query objects where class == [StopWatch] into sw
17
18     @snapshot s
19     match action start on sw into _, sw:
20         query objects \
21             where class == [Pendulum] into p
22
23     set half_cycles = 0
24     match group {0, inf}:
25         match property changed [elongation] on p where elongation == 0
26         set half_cycles = half_cycles + 1
27
28     match action stop on sw
29
30     set diff = abs(p.elongation - s:p.elongation)
31     if diff <= radians(3):
32         feedback success "Well done! " + \
33             "Your measurement lies well within the tolerance range."
34         if half_cycles / 2 >= 5:
35             feedback success "It was a good strategy to use " + \
36                 half_cycles / 2 + " cycles to improve the precision."
37     elif diff <= radians(5):
38         feedback warning "The measurement of the cycle duration could " \
39             "be a little more precise."
40     else:
41         feedback mistake "You did not measure a full cycle duration."
42
43     record
```

Fig. 7.5 Assessment rule used to capture the measurement of the cycle duration

The necessary assessment rule is depicted in Fig. 7.5. In line 16, all objects which are of class ‘StopWatch’ and currently known by the user, respectively the partial world model, are loaded into the variable ‘sw’. The ‘match action’ clause awaits a ‘start’ action on one of the available stopwatches. If a proper action is received the actually used object reference is stored (overwritten) in ‘sw’. As an inner condition (line 20), a proper ‘Pendulum’ object must be found and bound to variable ‘p’; otherwise, the pattern is aborted, as it makes no sense to measure a periodic cycle without any kind of oscillating system. If the action clause succeeds, a snapshot of the current environment state is taken, so that the pendulum can be compared with its previous state later. The following ‘match group’ clause will then repeat until the next ‘match action’ clause receives a stop action on the same stopwatch (‘sw’) as before. Prior to that, the nested ‘match property changed’ clause (line 25) catches all changes of the ‘elongation’ property of the pendulum, whenever the elongation passes zero. As an internal consequence, the count of half cycles will be increased. The primary consequences block starts at line 30, where procedural statements decide on the performance of the measurement and raise appropriate feedback messages. Future versions will also be able to log evidence for competence updates and mark tasks as completed.

```
pattern MeasurementSeriesLength:
    query objects where class == [Pendulum] limit 1 into p
    match pattern Measurement into m
    set mass = p.weight
    set count = 1
    match group {0, inf}:
        match pattern Measurement into m
        assert p.weight == mass
        set count = count + 1
    match first:
        match action submit
        match property changed [weight] on p
    if count > 4:
        feedback success "You have successfully performed a measurement " + \
            "series of " + count + " different lengths."
    elif count > 1:
        feedback warning "A measurement series should at least consist " + \
            "of 5 different lengths."
```

Fig. 7.6 Assessment rule used to detect a measurement series with different lengths

7.5.4 Systematic Measurement Series

After completing the simple introduction task, the student should investigate the dependencies that lead to different frequencies of the pendulum. Thus, it will be necessary to repeatedly measure the cycle duration of the pendulum while systematically changing a certain parameter. The parameters that can be changed are the mass of the bob and the length of the rope. But it is also possible to use different elongations, which will also result in different frequencies. According to an expert advice, this is often neglected by inexperienced learners.

In order to support the student, a behavior pattern is used (see Fig. 7.6) that will match against a group of multiple changes of a certain parameter—in this example the length of the rope—as well as a subsequent measurement of the cycle duration. The latter is implemented as a 'match pattern' clause, thus reusing the already existing behavior pattern for proper measurements. The group will be repeated as long as the mass of the bob stays constant. In addition, the number of repetitions with different rope lengths is counted as well. The moment any other parameter is changed, or the answer form is submitted, the series is considered finished. The primary feedback block will then evaluate how many measurements have been performed and appropriate feedback messages are triggered (see Fig. 7.7).

7.6 Challenges and Future Perspectives

Following the Horizon Report 2019 [7], it can be assumed that instruction is changing toward student-centered approaches, placing instructors into the role of facilitators and curators of learning. While emerging technologies—such as mixed reality approaches or adaptive environments—are expected to transform higher education within the next years, faculties and institutions will be required to support lecturers

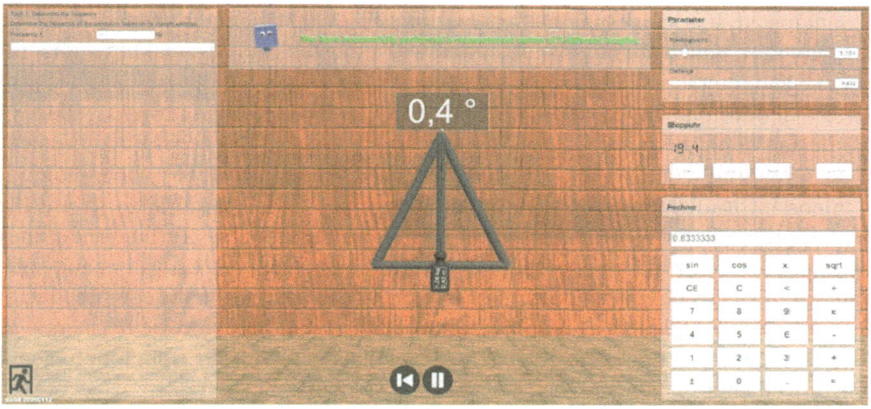

Fig. 7.7 The student receives positive feedback on completing a measurement series

with appropriate tools and expertise. In addition, the application of learning analytics will have an impact at all levels of higher education, including personalized adaption for individual students, as well as course and curriculum design.

Especially mobile learning has become the focus of content creation, which is related to the fact that mobile devices, such as smartphones, are owned by almost all undergraduate students and more than the half of adults worldwide [7]. This requires that content be responsive and can be synchronized with other devices, allowing learners to use their preferred one; also including virtual reality (VR), augmented reality (AR) and mixed reality (MR) approaches [7].

MR is anticipated as a relevant learning technology within the next two to three years as described in [7]. It subsumes different technological aspects of VR and AR. While the first term refers to entirely computer-generated environments with a high degree of immersion and interaction, the second describes the extension of the physical world with digital content. Both concepts can be realized with dedicated headsets or simple adapters for smartphones, although AR is to some degree also achievable with built-in smartphone technology alone. Benefits for learning are suggested in the context of experiential learning, which includes applications in inaccessible areas (remote places or unreachable locations), as well as simulated interactions with abstract concepts, usually found in science education. It is further possible to leverage such environments for authentic assessment scenarios [7]. However, institutions, which did not have the human resources for extensive individual student support in the past, are in our opinion less likely to make use of such opportunities in the future, as it will obviously be time-consuming without proper technological support for automation. This is where we see potential for an automated, flexible, behavior-oriented assessment approach as proposed in this chapter.

Besides, both technological concepts—VR and AR technology—are behind schedule according to the report [7]. Promising results exist, but there are still open issues and the expected scaling within the educational landscape has not been reached as anticipated. It is suggested that reasons might include the requirement of wearing huge goggles, which are not manufactured with adjustable lenses, making it uncomfortable and impractical for daily use, especially for wearers of glasses. But it is also noted that development is expensive, and expert knowledge on the proper pedagogical design could be missing. Another aspect is related to the general requirement of such learning content. It is emphasized that in order to adopt a technology, it is important to define the expected outcomes. Only then is it possible to use the most appropriate tool for the required learning setting [7]. This further underlines the importance of proper cooperation between domain experts, educators and software engineers. A flexible approach for interchangeable assignments combined with assessment and guidance, as proposed in our work, would provide the teachers—as domain experts and final pedagogical decision makers—with the right tool to adapt generalized environments to the needs of their students.

Finally, also artificial intelligent (AI) approaches are considered to significantly increase within the next few years [7]. While it is stated that certain skepticism exists, mainly due to ethical reasons such as privacy, equity and replacement of human workforce, potentials to implement better algorithms for adaptivity in personalized learning environments exist. This further includes virtual assistants based on natural language processing and speech recognition [7]. Due to the flexible and extensible nature of the Antares framework, future improvements such as pedagogical agents based on virtual assistants as well as improved adaptivity could be added without major changes. Once the feedback API is enhanced with audio capabilities, additions can be added to the evaluation engine without further changes to existing simulations.

7.7 Discussion and Future Work

Literature confirms the importance of immediate feedback and the necessity of guidance to enable successful learning, especially for beginners. While virtual reality and augmented reality are expected to shape the landscape of mobile learning within the next few years, there are still open issues such as cost-effectiveness, suitability for daily use and most notably the proper application of immersive environments in relation to the desired learning outcomes.

Several contributions to assessment and guidance in immersive environments exist. But most approaches lack flexibility in terms of platform-independence, reusability of content and assessment rules. Examples of flexible behavior-based assessment are limited to procedural skills, focusing only on predetermined sequences of user actions with slight variations and do not consider the situational context that is constituted by surrounding objects and their current state and behavior. In contrast, approaches on the other end of the spectrum rely on rather coarse metrics of basic actions or enclosed embedded tasks and do not consider the respective user actions in greater detail.

In this chapter, we have proposed the Antares framework, as a flexible, platform-independent and service-oriented approach to reusable behavior-based assessment that correlates the state of nearby objects with explicit user actions. This is achieved through an enhanced rule-based expert system that borrows aspects from regular expression pattern matching, data query languages and procedural languages. It further features exchangeable assignments through dynamically injected content items that may be used for instructions and background information as well as forms to submit results or implement traditional e-assessment items in-world. The Antares framework has been applied to a simple pendulum experiment placed inside an immersive 3D virtual laboratory. It requires the learner to measure the cycle duration and perform a systematic analysis of the influence factors that determine the frequency of the pendulum.

Although the first results are promising and demonstrate the technical feasibility of the approach, large-scale user studies are necessary to prove the increased usability and didactical advantages. This will require further collaboration with didactical experts, lecturers in higher education as well as school teachers. Beyond that, we are currently working on the implementation of a graphical editor for assessment rules, as well as an improved task structure to support a self-contained learning experience. In addition, the integration of richer semantics and the support of standardized ontologies is another direction we have already started following. Nevertheless, in the long term, advanced AI approaches should be investigated to further improve the concept, which might lead to—at least partially—automatically created assessment rules. Early contributions in literature demonstrate that data-driven approaches could be promising, also for vast open-ended immersive environments. Other aspects that will need further investigation include the assessment of collaborative workgroups, the proper integration of competence models to collect evidence of learning and provide suitable adaption, as well as learning analytics that offer subsequent reporting features for teachers as well as individual students.

Given that immersive approaches are still behind schedule, flexible, reliable and standardized assessment and feedback systems, that cover a wide range of assessment approaches, will be required for the efficient creation and application of proper learning settings that reach practitioners and teachers in the classroom.

Acknowledgements We want to thank Julian Wolf for implementing the new prototypes (pendulum, assessment interface) in Unity, respectively Maroon, as part of his master thesis. Special thanks to Ass. Prof. Dr. Johanna Pirker and her workgroup for providing us with the Maroon framework, especially to Michael Holly, who has updated and improved the prototypes whenever necessary. Parts of this research are further related to the Phantom3D project, which is dedicated to the improvement of physics education in large-scale university courses. It is internally funded at Graz University of Technology by the TEL Marketplace initiative, which promotes the implementation and usage of modern e-learning techniques. We also want to express our gratitude to all members and collaborators of the Phantom3D project team.

References

1. Carlile, O., & Jordan, A. (2005). It works in practice but will it work in theory? The theoretical underpinnings of pedagogy. *Emerging Issues in the Practice of University Learning and Teaching, 1,* 11–26.
2. Colburn, A. (2000). Constructivism: Science education's "grand unifying theory." *the Clearing House: A Journal of Educational Strategies, Issues and Ideas, 74*(1), 9–12.
3. Kirschner, P. A., Sweller, J., & Clark, R. E. (2006). Why minimal guidance during instruction does not work: An analysis of the failure of constructivist, discovery, problem-based, experiential, and inquiry-based teaching. *Educational Psychologist, 41*(2), 75–86.
4. Nicol, D., & Milligan, C. (2006). Rethinking technology-supported assessment practices in relation to the seven principles of good feedback practice. In Innovative assessment in higher education, pp. 64–77.
5. Spector, J. M. (2014). Conceptualizing the emerging field of smart learning environments.*Smart Learning Environments, 1*(1), 1–10. Retrieved from https://www.slejournal.com/content/pdf/s40561-014-0002-7.pdf.
6. Rutten, N., Van Joolingen, W. R., & Van Der Veen, J. T. (2012). The learning effects of computer simulations in science education. *Computers and Education, 58*(1), 136–153.
7. Alexander, B., Ashford-Rowe, K., Barajas-Murph, N., Dobbin, G., Knott, J., McCormack, M., Pomerantz, J., Seilhamer, R., & Weber, N. (2019). EDUCAUSE Horizon Report 2019 Higher Education Edition (pp. 3–41). EDU19.
8. Maderer, J., Pirker, J., & Gütl, C. (2018). Enhanced assessment approaches in immersive environments. In International Conference on Interactive Collaborative Learning (pp. 778–789). Springer, Cham.
9. Pirker, J., Lesjak, I., & Gütl, C. (2017). An educational physics laboratory in mobile versus room scale virtual reality–A comparative study. *International Journal of Online Engineering (iJOE), 13,* 106. https://doi.org/10.3991/ijoe.v13i08.7371
10. Morgado, L. (2013). Technology challenges of virtual worlds in education and training-research directions. In*2013 5th International Conference on Games and Virtual Worlds for Serious Applications (VS-GAMES)* (pp. 1–5). IEEE.
11. Crisp, G., Hillier, M., & Joarder, S. (2010). Assessing students in second life–some options. In*Proceedings ASCILITE Sydney Curriculum, technology and transformation for an unknown future,* pp 256–261.
12. Kickmeier-Rust, M. D., & Albert, D. (2010). Micro-adaptivity: Protecting immersion in didactically adaptive digital educational games.*Journal of Computer Assisted Learning, 26*(2), 95-105
13. Shute, V. J., Wang, L., Greiff, S., Zhao, W., & Moore, G. (2016). Measuring problem solving skills via stealth assessment in an engaging video game. *Computers in Human Behavior, 63,* 106–117.
14. Ma, W., Adesope, O. O., Nesbit, J. C., & Liu, Q. (2014). Intelligent tutoring systems and learning outcomes: A meta-analysis. *Journal of Educational Psychology, 106*(4), 901.
15. Mendjoge, N., Joshi, A. R., & Narvekar, M. (2016). Review of knowledge representation techniques for Intelligent Tutoring System. In*2016 3rd International Conference on Computing for Sustainable Global Development (INDIACom)* (pp. 2508–2512). IEEE.
16. Mitrovic, A. (2012). Fifteen years of constraint-based tutors: What we have achieved and where we are going. *User Modeling and User-Adapted Interaction, 22*(1–2), 39–72.
17. Aleven, V., Mclaren, B. M., Sewall, J., & Koedinger, K. R. (2009). A new paradigm for intelligent tutoring systems: Example-tracing tutors. *International Journal of Artificial Intelligence in Education, 19*(2), 105–154.
18. Floryan M., Dragon T., Basit N., Dragon S., Woolf B. (2015). Who Needs Help? Automating Student Assessment Within Exploratory Learning Environments. In Conati C., Heffernan N., Mitrovic A., Verdejo M. (eds)*Artificial Intelligence in Education.* AIED 2015. Lecture Notes in Computer Science, vol 9112. Springer, Cham

19. Gobert, J. D., Sao Pedro, M. A., Baker, R. S., Toto, E., & Montalvo, O. (2012). Leveraging educational data mining for real-time performance assessment of scientific inquiry skills within microworlds. *Journal of Educational Data Mining, 4*(1), 104–143.
20. Greuel, C., Myers, K., Denker, G., & Gervasio, M. (2016). Assessment and content authoring in semantic virtual environments. In*Proceedings of the Interservice/Industry Training, Simulation and Education Conference (I/ITSEC)*.
21. Greuel, C., Denker, G., & Myers, K. (2017). Semantic Instrumentation of Virtual Environments for Training. Retrieved from https://pdfs.semanticscholar.org/4758/97c345d490c35d86359d4051da551c71b5d0.pdf.
22. Fardinpour, A. (2016).*Taxonomy of Human Actions for Action-based Learning Assessment in Virtual Training Environments* (Doctoral dissertation, Curtin University).
23. Nowlan, N. S., Hartwick, P., & Arya, A. (2018). Skill assessment in virtual learning environments. In*2018 IEEE International Conference on Computational Intelligence and Virtual Environments for Measurement Systems and Applications (CIVEMSA)* (pp. 1–6). IEEE.
24. Schreiber N., Theyßen H., Schecker H. (2016). Process-Oriented and Product-Oriented Assessment of Experimental Skills in Physics: A Comparison. In: Papadouris N., Hadjigeorgiou A., Constantinou C. (eds) Insights from Research in Science Teaching and Learning. Contributions from Science Education Research, vol 2. Springer, Cham.
25. Hevner, A., & Chatterjee, S. (2010). Design science research in information systems. In*Design research in information systems* (pp. 9–22). Springer, Boston, MA.
26. Maderer, J., Gütl, C., & Al-Smadi, M. (2013). Formative assessment in immersive environments: a semantic approach to automated evaluation of user behavior in open wonderland. In*Proceedings of Immersive Education (iED) Summit*.
27. Maderer, J., & Gütl, C. (2013). Flexible automated assessment in 3D learning environments: Technical improvements and expert feedback. In E-iED 2013 Proceedings of the 3rd European Immersive Education Summit (pp. 100–110).
28. Buchanan, B. G., & Duda, R. O. (1983). Principles of rule-based expert systems. In Advances in computers (Vol. 22, pp. 163–216). Elsevier.
29. Friedl, J. E. (2006). Mastering regular expressions. " O'Reilly Media, Inc.".
30. Rybiński, H. (1987). On first-order-logic databases. *ACM Transactions on Database Systems (TODS), 12*(3), 325–349.

Chapter 8
Improving Electrical Engineering Students' Performance and Grading in Laboratorial Works Through Preparatory On-Line Quizzes

P. C. Oliveira, O. Constante, M. Alves, and F. Pereira

Abstract We have been implementing and evaluating a set of new teaching/assessment methodologies, in the context of a *circuit analysis* course of the *Electrical and Computer Engineering* degree at ISEP. One of them, reported in this chapter, is the implementation of on-line (Moodle) quizzes for students to prepare their lab classes in anticipation. We show that this methodology contributes to a more organized and consolidated preparatory work carried out by each student, as well as to an automatic (individual) grading along the semester. The qualitative feedback from the teachers' team (ourselves included) has been quite positive: students approach lab classes with more knowledge and confidence and finalize their lab scripts sooner and with better results, when comparing to the previous paradigm. This perception has been consolidated by a questionnaire-based study building on over 300 students/replies, confirming the students' preference for the new methodology. Importantly, lab *GPA* and *success rate* increased from 60 to 70% and 80 to 90%, respectively, when comparing to the old paradigm.

Keywords Electrical engineering education · Information and communication technologies (ICT) · Learning management systems (LMS) · Moodle · Automatic assessment and grading · Autonomous study · On-line quizzes · Electric circuit analysis

---

P. C. Oliveira (✉) · O. Constante · M. Alves · F. Pereira
Electrotechnic Department, ISEP, Rua Dr. António Bernardino de Almeida, 431, 4200-072 Porto, Portugal
e-mail: pco@isep.ipp.pt

O. Constante
e-mail: odc@isep.ipp.pt

M. Alves
e-mail: mjf@isep.ipp.pt

F. Pereira
e-mail: fdp@isep.ipp.pt

R. Babo et al. (eds.), *Workgroups eAssessment: Planning, Implementing and Analysing Frameworks*, Intelligent Systems Reference Library 199,
https://doi.org/10.1007/978-981-15-9908-8_8

8.1 Introduction

8.1.1 Context and Contribution

This chapter reports on our experience in implementing a new methodology for improving the students' autonomous work/skills and lab work preparation. The main objective of this methodology is to ensure that students carry out the autonomous work for the curricular unit in a more organized way and that the quality of that autonomous study is continuously assessed along the semester.

Students' autonomous work is stimulated through the proposal of preparatory exercises and simulations that must be done in anticipation of their laboratory classes. This has been implemented through Moodle quizzes, where automatic assessment enables students to understand where and why they failed. We can see this process as a cycle in which students carry out autonomous work (with guidance), are evaluated, check what they did wrong and repeat the process in new subjects, week after week. A preliminary insight of this (frame) work has been published and presented as an extended abstract, in Oliveira et al. [1].

In the "old" paradigm (read "in previous editions of the course") this preparation work was identified as "homework" (in each Lab Script, corresponding to each lab class). The problem was that only part of the students adequately addressed and completed their homework, even though it was highlighted as "mandatory", So, many students approached the lab class clearly underprepared, leading to obvious problems in completing their lab works correctly and timely.

As later illustrated in Fig. 8.1 and further elaborated, the average results of the Lab Assessment grading have clearly improved: GPA has increased from around 60% (old lab preparation paradigm—2015–2018) to over 70% (new Moodle-based preparation—beyond 2018) and the percentage of students above the threshold (eligible for the Exam) has increased from around 80% to over 90%.

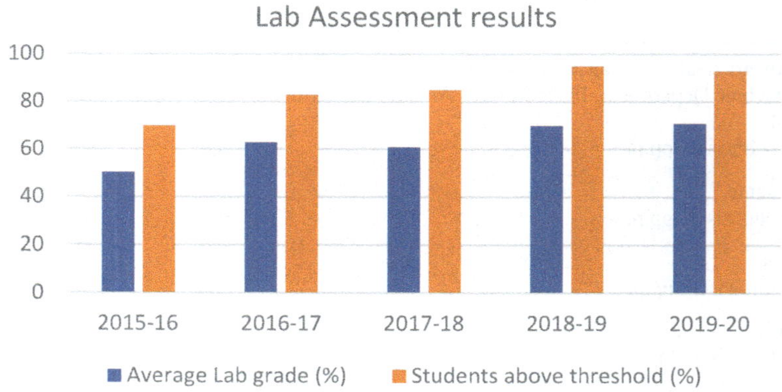

Fig. 8.1 Lab assessment results in the last editions of the FEELE course

The remainder of this chapter is organized as follows. In this introductory section, we contextualize the benefits of ICTs as tools for catalyzing autonomous work in higher education and look into the most relevant features of the Moodle Learning Management System, which has served as the platform to implement the lab preparation tests. Section 8.2 reports our case study, namely it describes the course at stake and its teaching and assessment methodology, as well as both "old" and "new" lab preparation paradigms. Section 8.3 digs into the evaluation of our new methodology, starting by identifying the research methodology and then elaborating on the presentation, analysis and discussion of the results of our study. Finally, Sect. 8.4 draws some conclusions and paves the way for future work.

8.1.2 ICT in Higher Education—An Overview

Information and Communication Technology (ICT) play a major role in our increasingly dynamic and highly technological society [2]. The recognition of the enormous potential of ICT-based tools for the social construction of knowledge and for autonomous and shared study enlightens us on the importance of a new culture: the digital culture. Creativity, competitiveness, and innovation are inherent characteristics of ICT, which evidence the development based on information and knowledge.

In education, the quest for a better and quick qualification of professionals for the labor market is increasing, together with an eagerness to attract and motivate students [3]. In what concerns Higher Education Institutions (HEIs), there is an additional challenge to encourage students to develop their individual knowledge skills in the most autonomous way and also to stimulate continuous and ongoing training.

According to Scotz and Ito [4], the various aspects related to the modernization of Higher Education (HE) involve the development of nation-wide assessment systems, the increase in the number of HEIs and the diversification of their modalities, as well as the use of new ICT for the improvement of teaching and research.

The application of ICT is creating significant changes in the teaching and learning process as it has several advantages over traditional teaching methods. As early as 1996, Smith [5] stated that ICT facilitates the immediate exchange of information, the adaptation of information to different learning styles and the encouragement of exploitation. According to Gredler [6], the integration of ICT helps in constructivist learning, in which students interact with other students, the teacher, information sources, and technology. ICT also provides tools that facilitate access to people, content, strategies, activities, guidance, and opportunities to apply new information that makes learning a personal process. Technology provides students with a choice of how, when, and where to participate in the learning process, and to bring together a variety of learning resources, including people, places, and materials they might never have access to [7].

Nowadays, students' needs are different as well as their habits. The use of computers, the Internet, and social networks changed the way students interact with the world. Several studies have been performed in order to characterize the habits of Internet/ICT usage by higher education students. One such studies, carried out in Portugal by Babo et al. [8], showed that most students access the Internet several times a day, that they are connected an average of 1–3 h per day and that the students who spend more time online are the ones that are enrolled in "technology" related courses. Moreover, the main reasons for students to use the Internet have been identified as [9]: (i) to research on work/study, (ii) accessing documents in their LMS (e.g., Moodle).

In this context, ICTs are widely used in various dimensions of higher education, either face-to-face or Distance Learning (EaD) through virtual learning environments (LMS), such as the Moodle platform—"Modular Object-Oriented Dynamic Learning Environment" [10].

Teachers play a crucial role in ICT integration. In 1998, Sarmento et al. [11] stated that "The widespread use of ICT by younger teachers is also a sign of confidence." Of course, with this new teaching management, and particularly in HE, teachers are faced with a new paradigm, since teaching today is not simply a transmission of knowledge [12]. Teachers soon started to interact in the construction of knowledge and become the drivers of research with new technologies [13]. According to the vast literature in the area, as a consequence of organizational, curricular, extracurricular, and political changes in HE, teachers are required to continuously acquire new skills. They must adapt supporting content objectively and clearly, making it attractive and enjoyable, but fundamentally educational-oriented, and all this process is quite challenging. However, many teachers do not take advantage of ICT potential to promote higher quality teaching/learning yet, even though they have a favorable attitude towards it [14, 15].

8.1.3 Student's Autonomous Study

The amount of autonomous study of HE students started to get more objectively and quantitatively clarified with the introduction of the *European Credit Transfer and Accumulation System* (ECTS) in 1989, within the *Erasmus* program, and later with the request from Bologna Ministers in the *Bucharest Communiqué* [16], in 2012. This was a Ministers' call on institutions to further link study credits with both learning outcomes and student workload and to include the attainment of learning outcomes in assessment procedures. According to the ECTS Users´ Guide [16], the workload is an estimate of the time the student typically needs to complete all learning activities (e.g., lectures, seminars, practical work, individual and group research, report writing, projects), and individual study required to achieve the defined learning outcomes.

Nowadays, a better estimate of the amount of time the student must allocate to autonomous work is available. Nevertheless, the quality of the autonomous work is difficult to quantify and to guarantee [17]. It is important to notice that now the student is the center of the educational process where he/she is expected to have and active and critical learning role. This is important, as we want the graduate student to be prepared to enter the job market, where this autonomy is required. On the other hand, it may be dangerous, as the student is left to his/her own way/risk without much help. In fact, the tendency is to simplify all this by decreasing the number of lectures and to "force" (read "motivate") the student to achieve the proposed outcomes through autonomous work.

Many students engaging HE get disoriented or even lost at some point and to some degree [18]. One of the reasons for this scenario is related to their educational background since very often the requirement for autonomous work at basic/secondary school is very low and when they access HE students face a new reality where autonomous work is preponderant [18]. Note that in HE the number of hours allocated to autonomous work is higher and the number of hours of contact with the teacher is lower than in high school.

Autonomous study implies the mobilization of many students skills: knowing the objectives that they have to achieve, knowing and recognizing what is being taught, knowing how to define/plan work tasks and priorities, knowing how to use countless information resources, knowing how to write summaries and prepare reading sheets and reports, knowing how to work in group, etc. Since rookie students do not have these autonomous work skills yet, it is paramount to find means of monitoring their evolution, i.e., the skills that students acquire during their study.

In this context, it is very important that teachers provide different kinds of supporting materials and tools to students, are available to answer their questions and regularly assess their (autonomous work) skills and knowledge along the way.

What do we really mean when we talk about students' autonomous study in HE? The answer to this question can be found in the following quotes. According to Candy [19], "Independent study is a process, a method and a philosophy of education whereby a learner acquires knowledge by his or her own efforts and develops the ability for enquiry and critical evaluation". Knight [20] defended that "Independence ... is not the absence of guidance, but the outcome of a process of learning that enables learners to work with such guidance as they wish to take ... getting there needs considerable insightful planning and action.". Thomas [21] states that "Broadly, independent learning is undertaken outside contact hours, but contributes to course-specific learning outcomes. ... Independent learning is undertaken by students, either on their own or with others, without direct input from teaching staff".

However, how do we know if students are carrying out good quality autonomous study? We do not want to reach the end of the semester and verify that the student failed to pass the curricular unit without identifying where/why he/she has failed. This is why it is extremely important for teachers to support and guide students throughout the semester. In addition, it is important to assess students' skills at different timings (and in different valences), so that the students have a relative and timely feedback regarding their working methodology and effort [22].

8.1.4 Learning Management Systems (LMS)

Engineering education faces new challenges, as XXI century students are different from those from the XX century and the teaching methodologies did not adapt and evolve accordingly, or at least not at an adequate pace. Students today are Internet-dependent and HEIs must leverage this fact to promote teaching and learning. The role of the teacher has changed (or should change), and other types of learning environments have been emerging, taking full advantage of ICT for teaching/learning. In this context, most HEIs have been using *LMS* for over a decade.

A LMS is a web-based software application that is designed to support learning content, students' interaction, assessment tools, reports of learning progress and all kind of student [23]. *Moodle*, *Sakai* and *Atutor* are among the most popular open-source LMS, while *Blackboard*, *SuccessFactors* and *SumTotal* are examples of their commercial counterparts. From these LMS platforms, our attention will fall on Moodle, since it is one of the most used LMS in higher education and the one used in our institution.

Moodle (*Modular Object-Oriented Dynamic Learning Environment*) was originally developed by Martin Dougiamas and available online for the first time in 2002 [24]. It is used in 230 countries by more than 176 million users, featuring over 20 million courses and 103 thousand sites [25]. In Portugal, there are 1246 registered sites [25]. Moodle enables to produce Internet-based courses and web content and it was designed to provide educators, administrators, and learners with a single robust, secure, and integrated system to create personalized learning environments and experiences. Importantly, it is open-source, which is quite appealing for higher-education institutions with ICT courses and programming competences such as our own (*Instituto Superior de Engenharia do Porto*—ISEP).

From the teacher's point of view, Moodle is easy to use (at a basic level) and has a plethora of functionality (please refer to Fig. 8.2). Basically, the interaction with Moodle can be divided into two main blocks: *Course Management* and *Resource/Activities* [26–28].

In *Course Management* you can perform all activities related to the formal part of the course, namely defining course settings, page layout, number of topics or weeks, enrolling students/teachers and assigning them roles, creating working groups, setting evaluation criteria, viewing student ratings, and producing all kind of reports.

The *Resources/Activities* are the key functionality for the interaction between teachers and students, as well as the direct interaction between students. In the *Resources*, the teachers can put all the supporting contents, such as slides, exercises, books, lab scripts, and exam examples. The *Activities* may include chats, forums, quizzes, lessons, or surveys.

Moodle has some interesting characteristics, namely [23, 29–31]:

- facilitates the distribution of content, as it is available online and can be accessed anywhere/anytime, which potentially increases students' motivation (specially for worker-students);

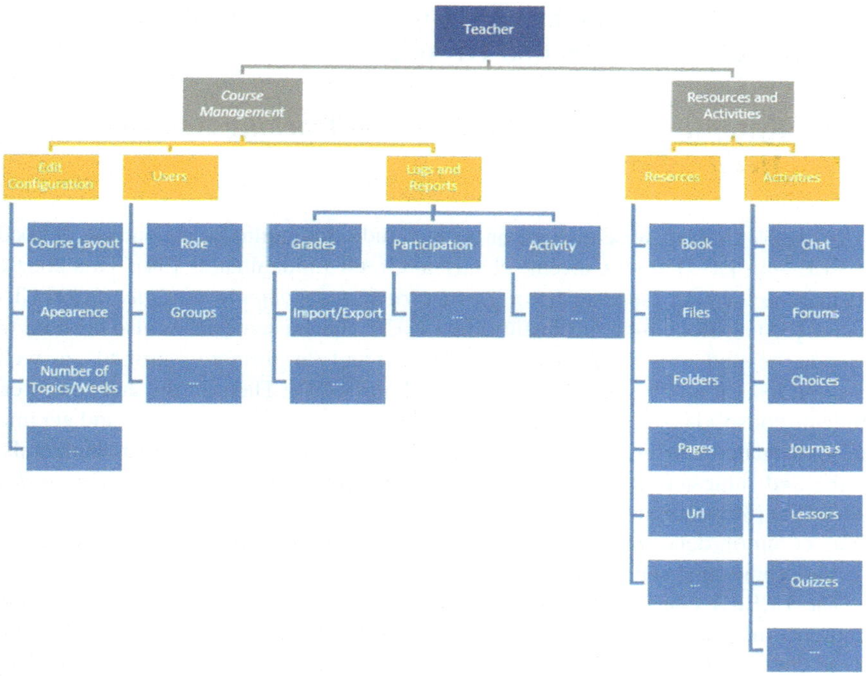

Fig. 8.2 Moodle: main blocks and functionality

- allows automatic students' assessment and grading, through tests/quizzes, enabling to provide them an immediate feedback;
- enables the teacher to more easily monitor students' activity (e.g., it is possible to check the records on when and how many times a student logged in);
- fosters a straightforward management of paper/report assignment submission and version control (e.g., by managing deadlines and file submission, timestamping, and storage).

Some studies indicate that the interactivity provided by Moodle tasks/assignments leads to more active students with a higher motivation and predisposition to learn [32]. Additionally, integrating online components into traditional classes has been shown to substantially improve the communication between students and teachers, to an increased access to Internet resources, and to a higher level of students' satisfaction [33].

8.2 Case Study—The FEELE Course and the Change in the Lab Preparation Paradigm

8.2.1 Synopsis of the FEELE Course—Teaching/Assessment Methodology

In this book chapter, we elaborate on a case study that builds on a course—dubbed FEELE—on DC (Direct Current) electrical circuit fundamentals [34]. This course fits into the 1st year, 1st semester of the *Electrical and Computer Engineering* (ECE) degree [35] curriculum at our institution—ISEP. The 1st semester is also devoted to provide the student with *Mathematics*, *Algebra* and *Programming* grounds, as well as some insight on *Project Management* and *Soft Skills*. Then, in the 2nd semester, students consolidate their knowledge in Mathematics and Programming, and get into AC (Alternated Current), Physics, Digital Systems and Electronics fundamentals. In the 2nd semester, the course on AC circuit fundamentals (TCIRC—*Teoria dos Circuitos*) [36] is aligned with FEELE (the focus of our study), and therefore shares some common denominators such as the core teaching team, teaching/assessment methodologies and self-learning tools.

The FEELE course introduces the most important concepts/terminology and theorems/methodologies for understanding and analyzing the behavior of DC electrical circuits, both in theory and practice. We start by the fundamental electrical quantities and the interrelation between them, such as Voltage, Current, Resistance/Conductance, and Resistivity (for different types of material), and Power/Energy. Then students learn methodologies to compute equivalent resistance and to apply voltage/current dividers. There is a module dedicated to measurement instruments and methods, introducing fundamental concepts such as measurement accuracy, errors and uncertainty, and the determination of measurement uncertainty and number of significant digits in measured quantities/computations (deriving from digital and analog instruments) and basics on the *Wheatstone Bridge* for resistance measurements. In the final module, the students get acquainted with different circuit analysis methods, e.g., namely the *Branch Currents Method* (based on the Kirchhoff Current and Voltage laws), the *Superposition Theorem Method*, the *Mesh-Current Method* (MCM, aka "*Loop-Current Method*"), the *Node-Voltage Method* (NVM), as well as simplification-oriented theorems such as *Thévenin*'s, *Norton*'s and *Millman*'s.

The teaching methodology of the FEELE course builds upon a (still too) "classical" approach of our institution (and most higher education institutions in Portugal, to our best knowledge), organizing lecturing into "Theoretical" (T-type), "Theoretical-Practical" (TP-type) and "Practical-Laboratorial" (PL-type) classes, as follows (lectures run for 15 weeks):

- **T-type** classes (2 h per week, up to 60–80 students in-class): these classes are more expositive, mostly building on slide show, although we have been reinforcing problem-based learning, through the analysis of practical examples and the resolution of exercises, providing a technological approach (whenever possible); we have also been running Kahoot-based quizzes, to grab students' attention and as an ice breaker during classes.

- **TP-type** classes (1 h per week, up to 30 students in-class): these classes are dedicated to the resolution of exercises, in synchronization with the topics that have been addressed in the T classes of the preceding week; the team has elaborated a series of exercise books, organized according to topics (e.g., *Equivalent Resistance, Kirchhoff Laws, Thévenin/Norton Theorems*), from which a subset is proposed for the students to work in each TP class; the teacher's role is to support students in the resolution of these exercises, clarifying any doubts and orienting them to take the best approach to each problem; the students are stimulated to support their analytical work with their scientific calculators and circuit simulators.
- **PL-type** classes (2 h per week, up to 18 students in-class): PL classes are devoted to laboratorial experiments, based on predefined lab scripts; each class/lab script comprises 3–4 different experiments under a certain "umbrella" (e.g., the second lab script addresses the study of the Ohm's Law and electrical resistivity and resistance in linear and non-linear components); typically, students team-up in groups of 3 around a workbench featuring basic test & measurement equipment (e.g., DC power source, digital and analog Multimeters, Galvanometer, Breadboard) and components (e.g., carbon resistors, potentiometers, decade resistance box, lamp).

These 5 h "in-class" per week correspond to a total of 75 h (the semester usually runs for 15 weeks). The students are also supposed to work "autonomously" for around 120 h (the FEELE course has been assigned 7 ECTS).

Towards a better consolidation of the students' learning process, we have been supporting the DC/AC circuit analysis teaching/learning process in a way that theoretical models are instantiated and validated in practice (through practical experiments in lab classes) and also through simulation (we recommend the QUCS open-source simulator). We incentivize students to use circuit simulators both for self-learning and also for preparing their lab classes a priori, by performing the theoretical (analytical) and simulation analysis of the circuits they are going to implement/experiment in the lab.

Concerning students' assessment, we have gradually been diversifying timing/types along the years, having recently converged to a methodology that seems to be the most adequate and balanced, as follows:

- **Lab Assessment** (50% of the final grade, 10/20 minimum threshold to pass and get eligible to the Exam):
  - Lab Test 1 (20% of the Lab Assessment grade): Individual test (theory and experiments), in the lab, in the middle of the semester.
  - Lab Test 2 (40% of the Lab Assessment grade): Individual test (theory and experiments), in the lab, at the end of the semester.
  - Lab Work (40% of the Lab Assessment grade):

(10%) Lab Works Preparation (Moodle)[1]: off-class, individual grading;
(20%) Lab Works Reports (Moodle): in-class, group grading;
(10%) Assiduity/Punctuality: in every lab class; individual grading.

- **Exam** (50% of the final grade, 8/20 minimum threshold to pass):
  - it covers all the subjects and is oriented for the resolution of exercises, sporadically with more theoretical/conceptual questions;
  - it has a duration of 90–120 min; students are supposed to use their scientific calculators but (usually) not allowed to consult any notes;
  - it is divided into three groups of questions, in turn subdivided by several alinea, to ease the organization in the student's mind and also the correction and grading.

8.2.2 Laboratory Preparation—From a Subjective to An Objective Paradigm

Lab Assessment Methodology

The Lab Assessment methodology builds on solid foundations, based on our experience and students' feedback, and obviously framed by school regulations:

- The Lab Assessment reflects 50% of the final grade and results from the work the student has carried out during all the semester and is intimately influenced by their lab-related knowledge and performance;
  - this grading is exclusively attributed by the student's lab teacher;
- A minimum threshold of 10/20 is required for the Lab Assessment; if a student gets lower than 10/20, he/she fails the course and cannot access the exam;
  - this requirement is to keep some pressure on the students since the beginning of the semester and to make sure that the student commits to prepare and execute all (or at least most) lab experiments;
  - we decided to fix this 10/20 threshold uniquely for the overall "Lab Assessment" grade and therefore not imposing any minimum thresholds for its components, i.e., Lab Test 1, Lab Test 2, and Lab Work have no individual thresholds; this is to guarantee that even if a student fails in one of the tests, he/she is able to balance that negative result with the other two components so that the weighted average is above the threshold;
  - in the last editions of the FEELE course, an average of 90–95% of all students succeeded and therefore have been eligible for exam, which we consider a very favorable scenario, considering that those 5–10% include students who quit

[1]This component is bolded to highlight that it is the focus of our case-study.

the course/ECE degree (due to different factors, such as changing to another degree/institution, lack of motivation, personal problems);

- A student which attends all lab classes (on time) gets 100% in the "Assiduity/Punctuality" component;
  - it should be pointed out that a subset (1–2) of all classes (16–18) run in the night shift, specifically scheduled for worker-students, and that sporadically these students miss classes (e.g., due to work in shifts or missions abroad);
  - worker-students have a special regime (according to school regulations) and they should not be (too much) penalized for not attending classes;
- The Lab Tests are scheduled for the middle (Lab Test 1) and end (Lab Test 2) of the semester;
  - they are done individually (with no consultation, except for the resistance color code), in a workbench with all the required equipment/components, and have an approximate duration of 50 min;
  - the students must be comfortable with circuit theory (solving problems and practical exercises) and implementation (interpreting a circuit diagram, implementing it and make the appropriate measurements and computations).

More specifically, in this chapter, our attention is devoted to the third component of the Lab Work assessment—**Lab Works Preparation**, which represents 1/10 of the final grade (1/4 of 40% of 10/20). In the following two sections, we outline the most relevant aspects of the paradigmatic change we have implemented.

Lab Preparation: The Old Paradigm—"Subjective" and "Narrow" Assessment

In previous editions of the FEELE course, this preparation work was identified as "homework" in the Lab Script for each Lab class. There is a total of 9–10 Lab scripts, corresponding to the effective number of classes where students execute lab experiments.

The lab scripts usually have an introductory section with fundamental theoretical background on the topics of that lab class. Then there are 3–4 sets of experiments, each of them defining a circuit diagram (to be analyzed/implemented), required equipment/components (e.g., DC Voltage Source, digital multimeter, galvanometer, breadboard), execution procedure (for instance: 1. *Implement the circuit in Figure*; 2. *Set the Voltage Source to 5 V*; 3. *Measure the voltage in resistance R4*) and questions to be answered (*Register the Voltage in R4, in V, with 2 decimals*: □□.□□).

The lab script "Preparation work" comprehended (and actually still does) the following tasks (that should be performed in this order):

1. Reading and understanding the theoretical grounds supporting the lab experiments;
2. Performing all the theoretical analysis to support the experiments, based on the corresponding fundaments, theorems, and methods;

3. Performing the corresponding simulations,[2] enabling the comparison between analytical, simulation and experimental results.

While this preparatory work was "mandatory", we actually did not have a formal procedure to check/assess if/when the student has done, and to which extent has done (and understood), all the preparatory "homework" (explicitly highlighted in the lab script). Our experience dictated that less than half of the students (roughly 1 out of (a team of) 3 students) adequately prepared the lab experiments. Often, we found students kind of lost, (at least) at the beginning of each lab class, some of them asking to perform/complete the "preparation" tasks in-class, which turns out to be unviable, due to time restrictions and considering that all members of the team should be concentrated in executing the lab experiments.

The only tangible way to check and assess all the homework would be to assess all the lab reports, for all teams, for all lab classes, which would not be scalable for the lab teacher. Each class generates 6 lab reports, which multiplied by 10 lab scripts would lead to 60 reports per class (throughout all the semester). Considering that each report has over 10 pages of theoretical and simulation analysis and experimental results and other verification/validation-related answers and that each professor may be teaching up to 6 classes (of 18 students, 6 teams), it is easy to understand that it would an herculean task to assess all lab reports. Moreover, this assessment would be "team-wise" (not "individual-wise"), since each team delivers one report per lab script, so all members of the team would get the same grade.

In this context, we opted for selecting 2 out of 10 reports to be assessed and averaging both grades. The 2 reports to be assessed were selected by the team of teachers and the students were only informed of this selection after they executed the lab scripts. We recognize the following problems with this "old" Preparation Work methodology/assessment:

- The assessment was based on a sample (1/5) of all lab works carried out by the students along the semester; from the total 10 lab scripts, the students were only assessed considering 2 of them, with all obvious implications.
- If a student failed a specific class (without an illegible justification) corresponding to the execution of one of the selected lab scripts, it would get a 0% grade. Note: if the student is granted a fault justification from the school board, he/she can execute the lab script off-class (specifically or in the Open Lab, which runs 2 h every Wednesday).
- The assessment was based on the "subjective" evaluation perspective of their lab teacher, i.e., each teacher assessed the lab reports of his/her lab students without any cross-correction involving other members of the teaching team. Of course, this subjective evaluation also has its good sides, as highlighted in the following section.
- There was no way to check/assess if the student has done his preparatory homework (before the beginning of the class); only upon students arrival to the lab

[2] We have been favoring the use of the *Quite Universal Circuit Simulator* (QUCS), due to its technical characteristics and to the fact that is freeware and open-source [37].

class the teacher could (hypothetically) check, team by team (for 6 teams), if the team (and not individually, element by element) had done the homework; moreover, it was obviously impossible for the teacher to make an appropriate check of all the homework items, considering his/her management duties, especially at the beginning of the class (e.g., checking students' assiduity/punctuality (one by one), briefing/warning students about important aspects of the lab experiments).

- The amount of (extra) paper produced, in terms of cost, for students (roughly 10 cents per page), for storage (regulations impose storing assessment-related material for at least 5 years), and for the environment (disposal after 5 years).

In this context, we have worked towards a more objective and heterogeneous assessment of the Lab works, which has been implemented through individual on-line quizzes, as outlined next.

Lab Preparation: The New Paradigm—"Objective" and "Wide" Assessment

Within the context of the FEELE (and TCIRC) courses, we have been exploring a set of new teaching/assessment methodologies, namely the implementation of on-line (via Moodle) quizzes where students must prepare their lab/experimental work classes in the week preceding the corresponding lab class. This has been implemented since the 2018–19 edition of the FEELE course (2 editions completed, to date).

Comparing to the previous paradigm, this new strategy for the "Lab Preparation" strongly contributes to a more serious/organized preparatory work carried out by each student, as well as enables an automatic (individual) grading. This perception has been confirmed by our evaluation study, In Sect. 8.3. Nevertheless, it should be pointed out that even before we carried out this questionnaire-based study, we already had got encouraging feedback from the teachers' team (ourselves included): students have been approaching the lab classes with more knowledge and confidence than in the past and finalize their lab scripts sooner and achieve better (correctness-wise) results.

Currently (since the 2018–19 edition of the FEELE course), students are individually and objectively assessed throughout all the 9–10 Lab Preparation instances, as Moodle quizzes are automatically corrected and graded. Then, the grades of all Moodle quizzes are averaged to get the overall Lab Preparation grade.

The average results of the Lab Assessment grading in the last editions of the FEELE course are presented in Fig. 8.1. The average *Lab Assessment* grade (detailed at the end of Sect. 8.2.1) is represented by blue bars (in percentage, for the sake of readability of the chart), while the percentage of *students above the threshold* (students with a Lab Assessment grade greater or equal to 10/20, therefore eligible for the Exam) is illustrated in orange bars.

FEELE editions 2015–16, 2016–17, and 2017–18 represent the "old" paradigm, where lab preparation was identified as "homework" in the Lab Script but was not quantitatively and individually assessed. In the most recent editions (2018–19 and 2019–20), students prepared their lab classes using Moodle quizzes (the new paradigm).

From Fig. 8.1, it can be observed that both the Lab Assessment average grade (blue bars) and percentage of students above the threshold (orange bars; students with grades greater or equal to 10/20, therefore eligible for the Exam) have increased in the editions implementing the Moodle-based Lab Preparation (2018–19 and 2019–20) when comparing to the "old" paradigm in the three previous editions.

Besides these positive results, we can still identify some disadvantages/limitations associated to the Moodle-based preparation of the lab classes, as elaborated next.

The "subjective" evaluation that has been mentioned as a con in the last section (deriving from the teacher's own evaluation/assessment perspective, context, timing) may also be considered as a positive factor, in specific situations, due to the human experience and flexibility to consider particular cases.

For instance, the teacher can analyze diagrams, schematics, and graphics, as well as occasional outliers deriving from wrong students' options (e.g., setting the voltage source to 10 V instead of 5 V) or to malfunctional equipment/components (that the teacher can consider/grade as correct or at least partially correct). The latter situation could be mitigated by introducing multiple options/numerical results in the answer, but this approach is most of the times not viable/scalable for coping with the myriad of scenarios that may occur.

We have implemented these on-line quizzes in Moodle, and therefore are limited by its functionality. For instance, we can no longer request students to draw things such as schematics (e.g., block diagram, circuit diagram), charts (e.g., the evolution of the resistance with current, in linear/non-linear components) or waveforms/screenshots (e.g., from an Oscilloscope), or to fill in a table (e.g., for an easier matching of analytical, simulation and experimental results), because Moodle cannot interpret/assess these formats. In this case, we may say that this change in "format" has an impact on the "content", i.e., the on-line quizzes sometimes imply a change in the way the questions are elaborated and in what we can actually ask the student to fill in/answer.

We still cannot guarantee the Lab Preparation is done individually. We are aware (and actually even stimulate) that some students gather physically together (or remotely interact via other means) to respond to these quizzes, either in a sporadic or regular basis. This limitation is hard to overcome, as we do not intend to force students into a room (under supervision) for carrying out this type of preparatory (weekly) assignment.

In spite of these disadvantages and limitations, overall we believe that this change in paradigm/methodology is positive, both from our (teachers') experience/feeling of students' behavior/performance in the lab, as well from the tangible improvement in the Lab Assessment average grades, when comparing to the previous edition of the FEELE course.

8.2.3 *Adapting Students' Lab Preparation to Moodle*

As already referred, the goal of each Moodle "Test" is to ensure that students are adequately prepared to perform laboratory work by understanding and applying fundamental concepts, terminology and methodology. Students must perform theoretical analysis and obtain analytical results for the same circuits that they will implement and experimentally analyze in the lab.

Each preparation test has an average of 12 questions and takes an estimated time of 40 min to complete (maximum one hour). All tests have automatic assessments and the assigned grade can be consulted by students after the end of each activity.

When accessing the Preparation Tests, the students are informed of the time they have to perform the test, which test they are making and some recommendations for the laboratory class. Teachers can access the number of students who have made the test and their grades (at any time). Figure 8.3 shows the beginning of one of these tests.

In order to understand if the students understand the theoretical concepts and to confirm the numerical results of the analysis they perform, the Lab Preparation quizzes include several multiple-choice and True/False questions. For each question, a database has been created, and every time a student accesses the quiz, a different True/False or multiple-choice question is generated. Figure 8.4 exemplifies one of these questions.

The Lab Preparation Test requires students to upload all circuit simulations for that lab script (using the QUCS—*Quite Universal Circuit Simulator*) in a specific field, as exemplified in Fig. 8.5.

This type of question is not "classified", because it would require extra work from each teacher. The main idea was just to access that question if any doubt arose with students' responses. To ensure that the simulation is correctly performed, a database of numerical questions has been created. After performing the simulation, students

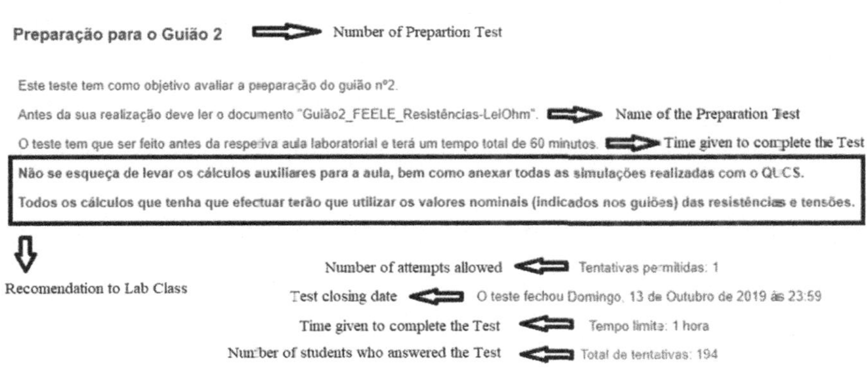

Fig. 8.3 Starting page of a Moodle test

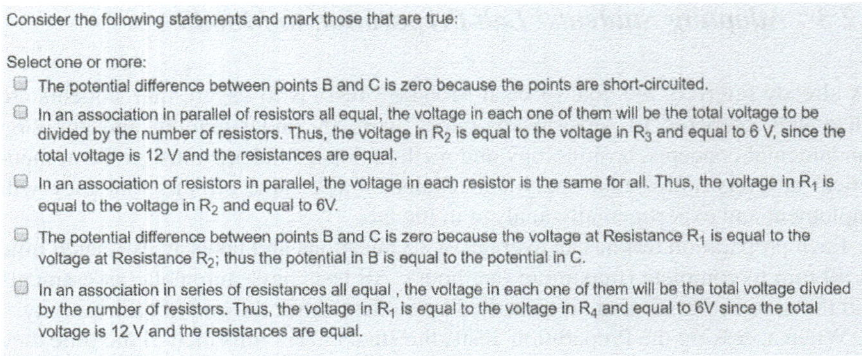

Fig. 8.4 Example of multiple-choice question

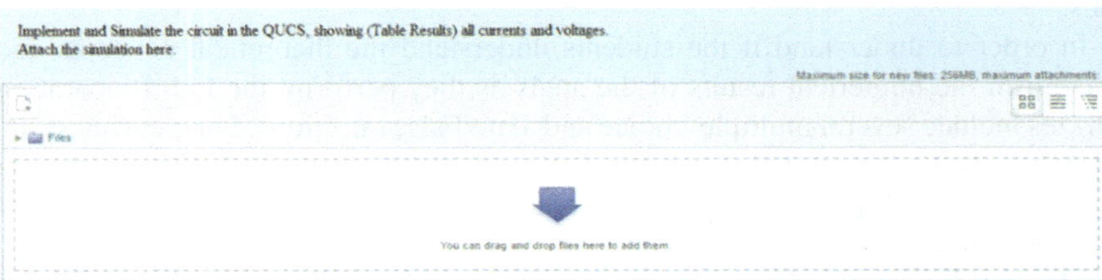

Fig. 8.5 Box to upload simulation files

have to answer one of these questions. An example of such type of questions is shown in Fig. 8.6.

In order to enable an automatic correction/grading, there was a need to adapt the questions related to graphical outputs. So, instead of students drawing the graph, they are asked to choose one out of a set of graphs. An example of this type of question is shown in Fig. 8.7.

In the old paradigm, the questions resulting from the theoretical analysis of electrical circuits, where different circuit variables (e.g., branch currents and node voltages) were calculated and output in the form of a table. Now, in the new paradigm,

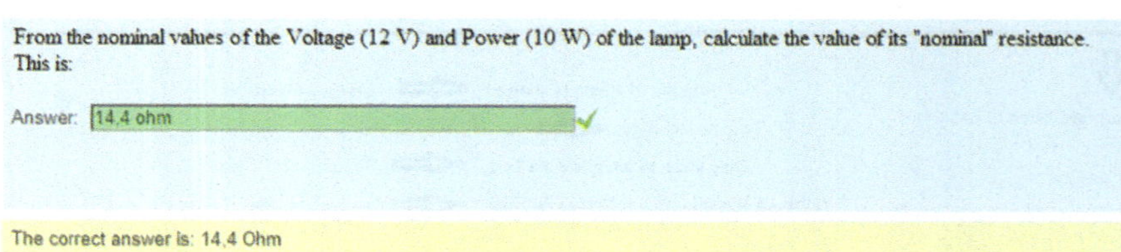

Fig. 8.6 Example of a numerical question to verify the simulation done by the student

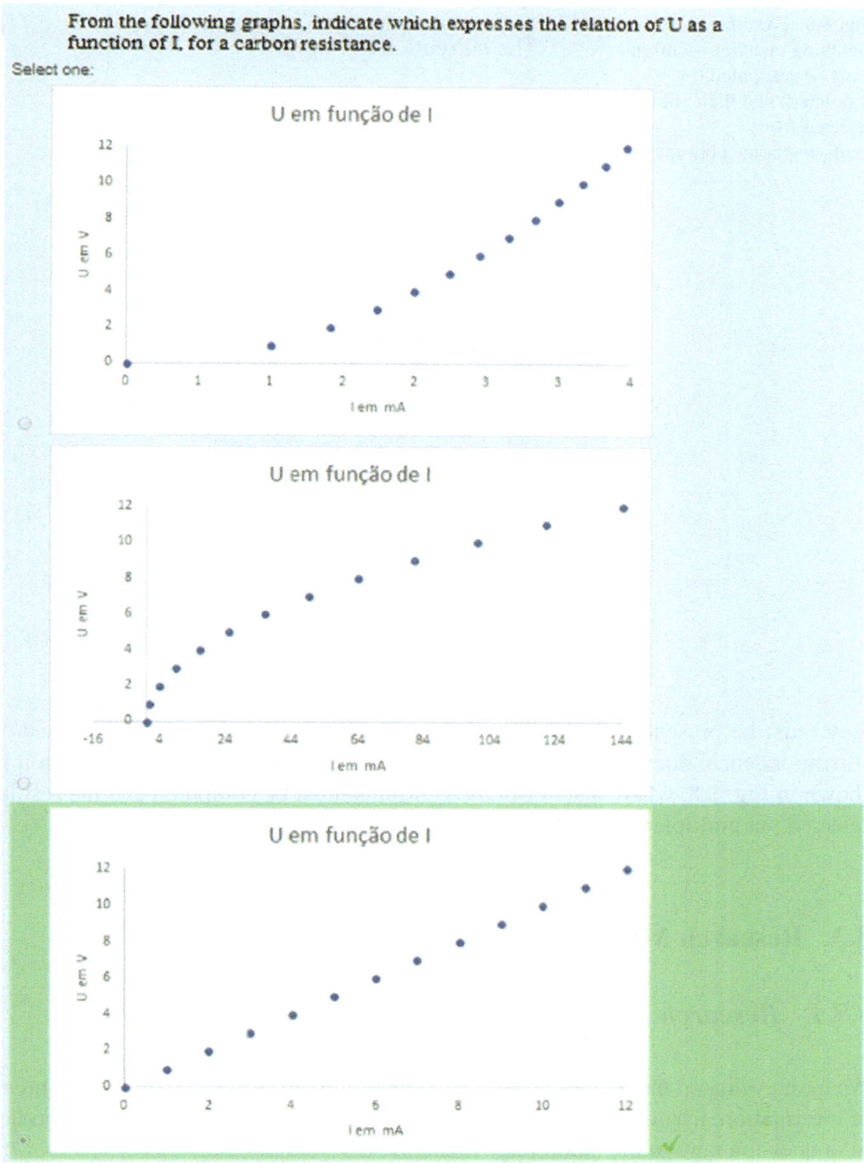

Fig. 8.7 Example of a question where students choose the correct option

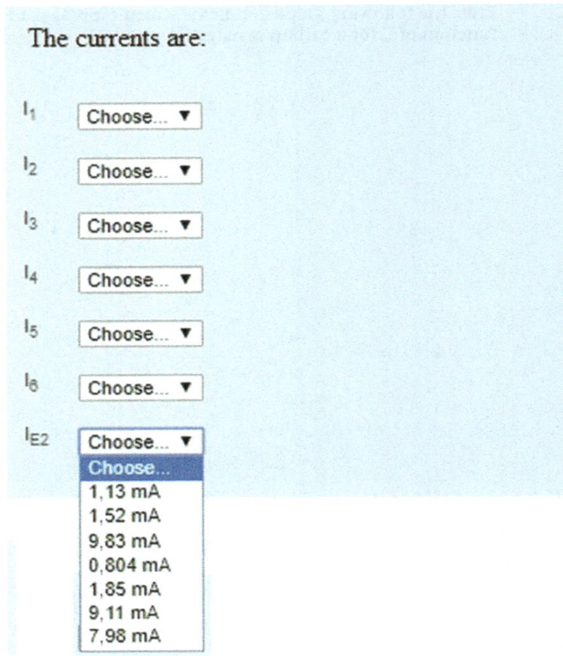

Fig. 8.8 Example of a matching question—currents must be computed (or simulated) and their values selected from multiple-choice pop-ups

they must be presented in the form of a sequence of numerical questions and/or correspondence, due to Moodle limitations. An example of this type of question is shown in Fig. 8.8, where a set of branch currents must be computed and the results selected via multiple-choice pop-ups.

8.3 Research Methodology and Evaluation

8.3.1 Research Methodology

We have evaluated the impact of the Lab Preparation quizzes based on three sources of information: i. a questionnaire to the students; ii. the reports produced by Moodle; iii. a questionnaire to the laboratory teachers.

The questionnaire to the students aimed at collecting the students' opinion on the implemented changes and the importance they attributed to them in their learning process, in the curricular unit of FEELE, namely in the preparation of the Laboratory Practice classes (PL) using Moodle.

The questionnaire featured 22 questions and was divided into two parts. The first part, with 4 questions, intended to characterize students in terms of age, gender, number of enrollments in the course unit, and number of hours (on average) that took to prepare the laboratory classes. The second part of the questionnaire was composed of 15 "closed" questions associated with a *Likert* scale (1–5, where 1 corresponded to "totally disagree" and 5 to "totally agree"). Finally, the last 3 questions were open questions about the positive and negative aspects that students found in these new methodologies implemented with Moodle.

The questionnaire was applied in the academic years 2018/2019 (1st year in which the preparation of the scripts in Moodle was implemented) and 2019/2020. The questionnaire applied in the academic year 2018/2019 had 4 extra questions for students who were repeating the course. These questions were intended to compare this new paradigm with the previous one. In the 2018/2019 academic year, 190 students answered the questionnaire, 50 of whom were repeating students; in the 2019/2020 academic year, 121 students responded.

All members of the FEELE teaching team were also asked to answer a specific questionnaire, divided into 2 parts with a total of 17 questions. The first part featured five questions that intended to teacher characterization (name, years of service, type of classes, and number of years teaching the FEELE course); the second part was composed of 15 closed questions associated with a Likert scale (1–5, 1 = "totally disagree" and 5 = "totally agree"). Finally, there were two open questions where teachers should indicate positive and negative aspects of the new paradigm.

The students questionnaires were analyzed using the SPSS—*Statistical Package for The Social Sciences*. The open questions were analyzed using NVivo, content analysis software.

8.3.2 Evaluation and Discussion

The 2018/2019 students questionnaires started to be analyzed in order to see if there would be statistical differences in the answers given by the students who attended the course for the first time and the repeating students (students who in the previous school year experienced the old paradigm). This analysis allowed us to conclude that in none of the questions (other than age) there were statistical differences in their answers. The same comparison was made for the students of the academic year 2019/2020 and, also, a non-statistical difference was found. In this way, we can analyze the students of each academic year (both rookie students and repeating students) as a single sample.

The next step was to make a descriptive analysis for the two academic years and, as mentioned above, differences in the responses of these two groups of students were sought, based on the *Mann–Whitney* hypothesis test for ordinary variables. From this first analysis, it was concluded that for most of the questions formulated there were no statistical differences between the two groups of students and, therefore, it was decided to analyze them altogether. For the four questions in which there are statistical differences between the two groups of students, the result of the Mann–Whitney Test is presented, and the answers were analyzed, separating the two academic years. The students average age of the sample (311 validated responses) is 20,1 years, with the youngest student being 17 and the oldest 64. Regarding gender, 11.3% are female and 88.7% are male. Regarding the frequency of the course, 80.7% attend the first time. Finally, the number of hours that students dedicate to preparing laboratory classes is 2.4 h (Fig. 8.9).

Table 8.1 shows, in the first 2 columns, the average value (μ) of the numerical classification given by students and the respective standard deviation (σ). The last 3 columns show the percentages below 3 (1 and 2), equal to 3 and above 3 (4 and 5) on the Likert scale. Levels 1 and 2 were grouped, as being below the neutral value (3) they are the equivalent to students' disagreement in relation to what was asked. Levels 4 and 5 were also grouped, as being above the neutral value means the students' agreement in relation to what was asked.

Analyzing Table 8.1, the preparation of scripts in Moodle was well accepted by most students, since the average value in all questions was well above the average value of the Likert scale.

Regarding question 1, it can be said that in the academic year 2019/2020 there was a greater number of students (94% against 82% from previous year) who consider having been informed of the importance of preparing the scripts. This value can be explained for two reasons: (i) students of the 2018/2019 school year were less aware of this new paradigm compared to the existing one in previous school years; (ii)

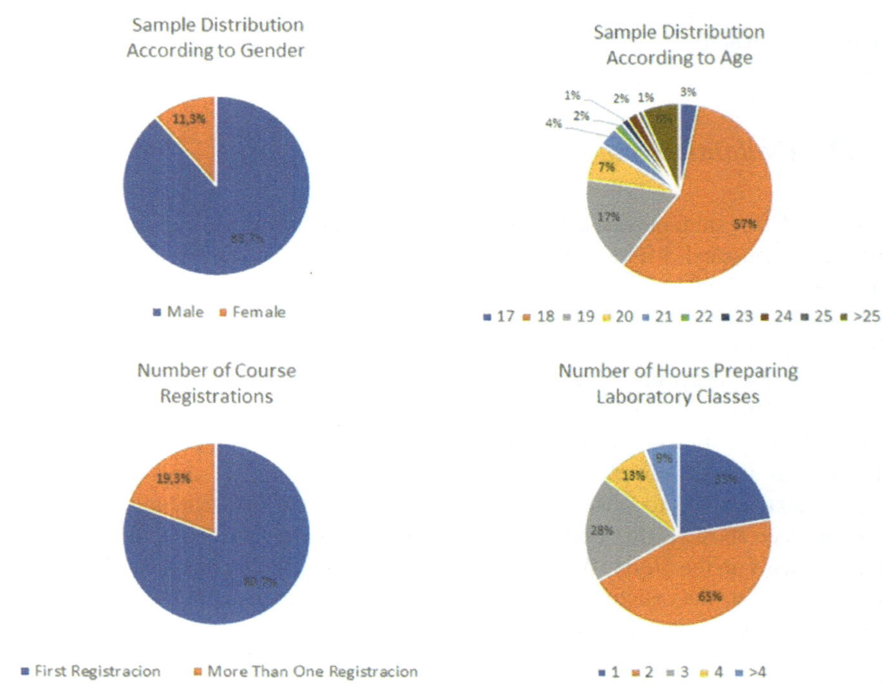

Fig. 8.9 Sample distribution according to gender, age, number of registration in the course and number of hours to prepare laboratory classes

Table 8.1 Summary of the students' feedback on the use of Moodle quizzes for the preparation of lab classes (311 validated responses)

Question		Average (μ)	Standard deviation (σ)	Likert Scale Values		
				< 3	=3	>3
1 I was informed about the need to prepare PL classes in Moodle ($W = 21{,}218{,}5$; $p = 0.000$)	2018/19	**4.27**	0.821	2%	16%	**82%**
	2019/20	**4.64**	0.649	1.7%	4.3%	**94%**
2. I usually prepare PL classes before they take place		**4.39**	0.835	3.5%	10.3%	**86.2%**
3. The preparation of PL classes in Moodle helps me to understand the work that I will do ($W = 20{,}252{,}5$; $p = 0.003$)	2018/19	**4.34**	0.759	2%	9.8%	**88.2%**
	2019/20	**4.56**	0.725	1.7%	3.4%	**94.9%**
4. It is easier to prepare laboratory scripts using Moodle than in a traditional way (read the script and make the proposed TPC's) ($W = 20{,}174$; $p = 0.005$)	2018/19	**4.20**	0.962	6.2%	16.5%	**77.3%**
	2019/20	**4.45**	0.924	4.3%	9.4%	**86.3%**
5. The preparation of scripts in Moodle helps me in learning the various contents taught in FEELE		**4.41**	0.739	1.6%	7.4%	**91%**
6. I prefer to prepare PL laboratory classes in Moodle		**4.26**	0.950	5.8%	11.3%	**82.9%**

(continued)

Table 8.1 (continued)

Question		Average (μ)	Standard deviation (σ)	Likert Scale Values		
				< 3	=3	>3
7. The sequence of questions in the tests of preparation of laboratory work in Moodle is in accordance with the sequence of work that will be performed ($W = 19{,}750{,}5$; $p = 0.029$)	2018/19	**4.36**	0.686	1%	8.8%	**90.2%**
	2019/20	**4.51**	0.690	0.9%	6%	**93.1%**

sharing the experience of students who had already attended (in the previous school year) this new evaluation model made the new students be more aware (2019/2020).

Regarding the preparation of the lab scripts in Moodle, there are no statistical differences between the two academic years. About 86.2% of the students stated that they had prepared the scripts with regularity. This value is in accordance with the information collected in Moodle, where (on average) 76.6% of students prepared all laboratory scripts. The 10% difference between these values can be explained by the fact that some students do not attend all lab classes along the semester; actually, some of them give-up the course/degree during the semester. Note that the students who answered the questionnaire are students who attended lab classes until the end of the school year while the percentage calculated via Moodle is made in relation to the total number of students enrolled (whether or not they attended the course).

Analyzing question 5, it can be stated that 91% of students recognize the importance of preparing scripts in learning the concepts of the course. Regarding question 6, 82.9% of students prefer preparation via Moodle over the traditional paper-based method.

In questions 4 and 7, although there are statistical differences between the answers given by the students in the two academic years, it can be inferred that the two groups are mostly of opinion that:

- it is easier to prepare laboratory scripts using Moodle (77.3 and 86.3% for the academic years 2018/2019 and 2019/2020, respectively);
- the sequence of questions asked in the preparation of the scripts is in accordance with the sequence of execution of the laboratory scripts (over 90% of the students, in both years).

As mentioned in the previous paragraphs, the questionnaire applied in the academic year 2018/2019 had 4 extra questions exclusively for students who were

Table 8.2 Summary of repeating students' feedback about the previous academic year with the old paradigm of paper-based lab preparation

Question	Average (μ)	Standard Deviation (σ)	Likert scale values		
			< 3	=3	>3
1. In the previous academic year (s) in which I attended FEELE, I always prepared the scripts	**3.31**	1.137	20.7%	**48.3%**	31%
2. In the previous academic year (s) in which I attended FEELE, I prepared the scripts individually	**2.76**	1.354	34.5%	**41.4%**	24.1%
3. In the previous academic year (s) in which I attended FEELE, I prepared the scripts in group	**2.93**	1.412	29.6%	**40.7%**	29.6%
4. Compared to previous years, the preparation of the lab works in Moodle helps me to better understand the works	**3.00**	1.414	20.7%	**48.3%**	31%

repeating the course. Table 8.2 shows the responses of these students to these questions.

Analyzing the questions, it is noted that there are many students who choose 3 on the Likert scale. This option may be justified by certain indecision in the option to be taken. In question 1, only 31% of the students confirm that always prepared the laboratory scripts. This number is in line with the feedback from the teachers, in which they stated that roughly just half of the students effectively prepared the laboratory scripts.

Analyzing the responses to questions 2 and 3, we can conclude that students prepared the laboratories both individually and in groups. Regarding question 4, it once again shows the indecision level of these students, since the highest response rate is at level 3 of the Likert scale.

Finally, the open questions aimed at obtaining the students' opinions on the positive and negative aspects of this strategy. Most students mentioned as positive aspects:

- *helps to understand the work to be done in the lab class*;
- *"compels" to study every week*;
- *allows detailed calculation of the analytical part of the laboratory work*;
- *digital support facilitates the evaluation.*

As negative aspects, the students indicated the possibility of the internet failure and of not being able to immediately have the correct answer if they make a mistake.

All members of the FEELE teaching team were also asked to answer a specific questionnaire. Regarding the teacher questionnaire, there were 6 responses. The number of years teaching in Higher Education is between 14 and 22 and all of them teach laboratory classes. The teachers were unanimous regarding the following aspects:

- more students are preparing the lab scripts;
- students are better prepared to do the laboratory work;
- students were more motivated to laboratory classes;
- classes are more productive and have greater dynamics;
- they prefer this new assessment system regarding the preparation of laboratory scripts compared with the previous one.

When asked about positive aspects about the script's preparation in Moodle, 4 teachers mentioned that a higher percentage of students are more aware about the laboratory work. Quoting one of the teachers: "*In this way, we make sure that all students prepare the script, or at least look at the subjects and type of experiences that will be addressed*".

Regarding the negative aspects, 4 teachers referred that it is impossible to ensure that it was the student who made the preparation of the script. Quoting one of the teachers: "It can favor students who do not study because they can have many right answers at the expense of sharing the results on social networks."

8.4 Conclusions and Future Work

Every year a new pool of students engages in the Electrical and Computer Engineering course at ISEP. The main (students) stream is ranked according to a weighted average of grades in high-school (10–12th scholar years) and specific exams in mathematics and physics, obviously limited to the *numerus clausus*. We are talking about over 150 new students arriving from this core contingent, the vast majority of them being 17–18 years old.

Since 2018–19, we need to restrain our surprise to see these rookies have been born after the year 2000.

This new generation of students is fully embedded in the digital world since birth, where the Internet, mobile platforms, social networks, and anytime/anywhere access to an exponentially increasing amount of information and sources make part of their *modus vivendi*. Useless to state that innumerous sources of distraction arise, both in-class and in their daily lives, in what respects the focus, energy an time we expect from them for attaining the necessary scientific, technological and soft skills and matureness to integrate companies or to leverage their own entrepreneurship just after 3 years of higher education.

On the other hand, the teaching staff, especially considering tenure teachers (the authors of this paper included), is getting older, which inflates the generation gap by the year.

Bottom-line, attracting good students to Engineering and ICT-related courses (that require mathematics/physics/programming skills that are typically scary for many youngsters), captivating them to attend and be productive in every lecture, motivating them to study and learn by their own, and to find the resilience to finalize their studies (around 10% give up), are very challenging and complex tasks.

Therefore, teachers need to find strategies to bridge this ever-increasing gap and turn the higher education experience the most successful, on behalf of the students and of the (hosting) society. It is not enough for higher education teachers to be constantly studying and updating the supporting materials, not even the ones driving scientific and technological research and transferring that knowledge "outdoor" (with/to companies) and "in-door" (feeding the courses/students). New teaching methodologies must be found, to encompass all the particularities of this (new) generation of (new) students. We cannot just say that "they are wrong", "they have bad habits", "they do not make any effort", "they are always distracted in-class", "they do not work enough by their own"; we need to find new ways of teaching, motivating and giving them tools for stimulating their self-learning skills.

Towards this objective, in the context of the electrical circuit analysis courses under our responsibility (FEELE and TCIRC, as identified at the beginning of Sect. 8.2.1), we have been implementing "traditional" strategies such as problem-based learning, reinforcing the technological flavor of typically "scientific/theoretical" courses, building knowledge through scientific (analytical/probabilistic), simulation and experimental models, and laboratorial/practice-driven learning. More recently, we have also been designing self-learning/teaching tools based on circuit simulators (both commercial-off-the-shelf such as QUCS and custom-made) and in-class quizzes/gamification (e.g., based on Kahoot or Socrative).

Within this framework and mindset, this chapter outlines another line of work that is oriented towards stimulating students to prepare their lab classes/experiments in advance and to enable their automatic and individual assessment in this task. This has been implemented through on-line quizzes in Moodle, which is the official *LMS* used in our institution. We have redesigned the way students' lab-related homework is done by readapting/reinventing the previously used lab scripts in a way that preparatory work (that must be done in anticipation to the corresponding lab class) has been separated from the actual lab script (executed in the lab).

As soon as we have started implementing this new methodology (first semester 2018–19), our feeling was that it was actually working, in the sense that we felt the students better prepared for executing the lab experiments and that they completed the lab scripts sooner and with more accurate results. However, this subjective feeling only materialized after the systematic and objective study that we have carried out and is reported in this chapter. The overall Lab Assessment GPA has increased from around 60% (old lab preparation paradigm—2015–2018) to over 70% (Moodle-based preparation—after 2018) and the percentage of students above the threshold (to be eligible for the Exam) has increased from around 80% to over 90%. Moreover, from over 300 answers to a special-purpose questionnaire, most respondents (roughly 80%, considering different questions/dimensions) feel enthusiastic about this new lab preparation/assessment methodology, including the students repeating the course (that have experienced the "old" paradigm).

In summary, and as a concluding remark, we are confident this is the way to go and therefore more time and effort should be dedicated to extend these strategies for stimulating students' self-learning to other dimensions and to other courses. In this line, we have also been conceiving Moodle quizzes for executing the lab scripts/reports (in-class, in real-time). Although we are still collecting data, the first impressions are quite positive, since students concentrate more in the lab experiments and do not waste (so much) time writing the report (on paper). This new approach also enables the automatic correction/grading of the lab reports, reducing the teachers' burden and guaranteeing an even assessment over all lab reports. Nevertheless, all these implementations must be improved, and the results must be validated with more data and statistical analysis.

References

1. Oliveira, P. C., Constante, O., Alves, M., & Pereira, F. (2019). Boosting students' preparation and automating gradings for laboratory classes via Moodle quizzes. In *CASHE, New Techological Appraches in the Educational Praxis of Higher Education* (p. 58).
2. UNESCO. (2002). *Information and communication technology in education: A curriculum for schools and programme of teacher development.*
3. Tinio, V. L. (2003). *ICT in Education: UN Development Programme.* Available: www.eprmers.org.
4. Scoz, B. J. L., & Ito, M. C. R. (2013). Ensino Superior e psicopedagogia: A busca por uma graduação alinhada com a contemporaneidade. *Rev. Psicopedag., 30*(91), 74.
5. Smith, K. L. (1996). Preparing faculty for instructional Technology: From education to development to creative independence. In *CAUSE Annual Conference: Broadening Our Horizons: Information, Services, Technology.*
6. Gredler, M. (2000). *Learning and instruction: Theory into practice.* New York: Prentice Hall.
7. New Media Consortium. (2007). The Horizon report: 2007 edition. Available: https://www.nmc.org/pdf/2007_Horizon_Report.pdf.
8. Babo, R., Rodrigues, A., Lopes, C., Oliveira, P. C., Pinto, M., & Queirós, R. (2011). Differences In internet and LMS usage—A case study in higher education. In R. B. Azevedo (Ed.), *Higher education institutions and learning management systems: Adoption and standardization.* Hershey: IGI Global.
9. Babo, R., Teixeira Lopes, C., Rodrigues, A., Pinto, M., Queirós, R., & Oliveira, P. C. (2010). Comparison of Internet usage habits in two generations of higher education students—A case study. In *Second International Conference on Computer Supported Education* (pp. 415–418).
10. Moodle. Available: https://moodle.org/.
11. Sarmento, M. J., Sousa, T. B., & Ferreira, F. I. (1998). *Tradição e Mudança na Escola Rural.* Lisboa: Ministério da Educação.

12. Angadi, G. R. (2014). An effective use of ICT is a change agent for education. *Online International Interdisciplinary Research Journal, 4,* 516–528.
13. Zhao, Y., & Cziko, G. A. (2001). Teacher adoption of technology: A perceptual control theory perspective. *Journal of Technology and Teacher Education, 9*(1), 5.
14. Barolli, E., Bushati, J., & Karamani, M. (2012). Factors that influence in the adoption of ICT in education. In *International conference on educational sciences, challenges and quality development in higher education.*
15. Cubukcuoglu, B. (2013). Factors enabling the use of technology in subject teaching. *International Journal of Education and Development Using ICT, 9*(3).
16. Bucharest Communiqué. (2012). Available: https://ec.europa.eu/education/ects/users-guide/docs/ects-users-guide_en pdf.
17. Holmes, A. G. (2018). Problems with assessing student autonomy in higher education, An alternative perspective and a role for mentoring. *Educational Process: International Journal, 7*(1), 24–38.
18. Neri de Souza, D. (2006). *Procedências dos Alunos e o Sucesso Académico. Um Estudo com Alunos de Cálculo I e Elementos de Física da Universidade de Aveiro.* Aveiro.: Universidade de Aveiro.
19. Candy, P. C. (1991). *Self-direction for lifelong learning: A comprehensive guide to theory and practice.* San Francisco: Jossey Bas.
20. Knight, P. (1996). Independent study, independent studies and 'core skills' in higher education. In *The Management of Independent Learning.*
21. Thomas, E. (2014). *Effective practice in independent learning.* Available: https://www.lizthomasassociates.co.uk/ind_learning.html.
22. Young, J. R. (2002). Homework? What homework? Students seem to be spending less time studying than they used to. *The Chronicle of Higher Education.*
23. Kasim, N. N. M., & Khalid, F. (2016). Choosing the right learning management system (LMS) for the higher education institution context: A systematic review. *International Journal of Emerging Technologies in Learning, 11*(6).
24. Grant, B., Samos, S., Hoare, S., & Torres, L. (2018). Measuring the success of Moodle at the University of Belize, Belize City Campus. In *Second Annual Research for National Development Conference.*
25. https://stats.moodle.org/. Accessed in December 2019.
26. Nash, S. S., & Moore, M. (2014). *Moodle course design best practices.* Birmingham: Packt Publishing.
27. Henrick, G., & Holland, K. (2015). *Moodle administration essentials.* Birmingham: Packt Publishing.
28. Büchner, A. (2016). *Moodle 3 Administration.* Birmingham: Packt Publishing.
29. Lustek, A., Jedrinovic, S., & Rugelj, J. (2019). Supporting teachers in higher education for didactic use of the learning environment Moodle. In *SLET-2019—International Scientific Conference Innovative Approaches to the Application of Digital Technologies in Education and Research.*
30. Susana, O., Juanjo, M., Eva, T., & Ana, I. (2015). Improving graduate students' learning through the use of Moodle. *Educational Research and Reviews, 10*(5), 604–614.
31. Meikleham, A., & Hugo, R. (2018). Understanding informal feedback to improve online course design. *The European Journal of Engineering Education, 45*, 4–21.
32. González, A. B., Rodríguez, M. J., Olmos, S., Borham, M., & García., F. (2013). Experimental evaluation of the impact of B-learning methodologies on engineering students in Spain. *Computers in Human Behavior, 29*(2), 370–377.
33. Chung, C., & Ackerman, D. (2015). Student reactions to classroom management technology: Learning styles and attitudes toward moodle. *Journal of Education for Business, 90*, 217–223.
34. Fundamentos da Engenharia Electrotécnica—FEELE (Electrical Engineering Fundamentals). *1st semester of the 1st year course on DC circuit analysis, of the Electrical and Computer Engineering (ECE) degree.* Instituto Superior de Engenharia do Porto (ISEP).

35. Oficial web site of ISEP's Electrical and Computer Engineering degree. *Licenciatura em Engenharia Electrotécnica e de Computadores*. Available: https://www.isep.ipp.pt/Course/Course/23.
36. Teoria dos Circuitos—TCIRC (Circuit Theory), 2nd semester of the 1st year course on AC circuit analysis. Electrical and Computer Engineering (ECE) degree, Instituto Superior de Engenharia do Porto (ISEP).
37. QUCS. Available: https://qucs.sourceforge.net/.

Chapter 9
Actively Involving Students by Formative eAssessment: Students Generate and Comment on E-exam Questions

U. Niederländer and E. Katzlinger

Abstract A reported problem of online learning scenarios is sustaining students' activity and motivation. In order to actively involve the learners as well as to increase their reflection on the learning content, students create multiple choice exam questions in the field of Digital Business Law in an online learning environment and mutually assess each other´s work by giving feedback in a peer review learning setting. In this process, students developed questions and provided comments on questions handed in by their fellow students. This learning scenario should lead to a better understanding of the learning content, facilitate collaborative learning and increase the students' assessment skills by addressing higher cognitive levels (as seen in Bloom´s taxonomy). Moreover, the aim was to enhance their contribution to the learning and teaching process in a more active way. Generating e-exam questions is part of the blended learning scenario MUSSS (Multimedia Study Services Social Sciences and Economics) offered at the Johannes Kepler University in Austria, in which the module "Digital Business Law", mandatory in the master's program "Digital Business Management", is integrated. The paper at hand reports on this learning scenario and its experiences, accompanied by a quantitative study (N = 34).

Keywords eAssessment · Peer learning · Higher education · Generating e-exam question · Blended learning

U. Niederländer · E. Katzlinger (✉)
Institute of Digital Business, Johannes Kepler University, Altenberger Strasse 69, 4040 Linz, Austria
e-mail: elisabeth.katzlinger@jku.at

U. Niederländer
e-mail: ursula.niederlaender@jku.at

R. Babo et al. (eds.), *Workgroups eAssessment: Planning, Implementing and Analysing Frameworks*, Intelligent Systems Reference Library 199,
https://doi.org/10.1007/978-981-15-9908-8_9

9.1 Introduction

9.1.1 Background

A high number of students at the university work while studying [1] and therefore require a learning environment that allows flexible, spatially independent learning. In this area, in particular, digital learning offers advantages over traditional on-campus courses. However, not only students who are online learners profit from digital learning activities, but also university face-to-face teaching is enriched by digital learning environments that enable and facilitate interactive, communicative and cooperative learning. Correspondingly, offering digital content may also lead to cost savings and/or an improvement in teaching quality compared with traditional university teaching methods [2].

Since there was a growing demand with regard to flexible e-learning scenarios due to a high number of students being forced to work at the Johannes Kepler University Linz in Upper Austria [3],[1] the blended learning program MUSSS (Multimedia Studies Services Social Sciences and Economics) was launched at the Faculty of Social Sciences, Economics and Business (SOWI) 11 years ago. The objective was to improve the learning environment through digital courses. The program's vision with regard to the didactic concept was to implement a technology-enhanced parallel program to the existing on-campus courses, appealing mainly to part-time professional students or those taking care of children. In the scope of MUSSS, courses in the field of economics and business administration are offered as both face-to-face and online courses [4]. Two-thirds of the bachelor curriculum for Business and Economics is held solely in the form of online or blended learning courses, whereas one third of the courses are realized on-campus in order to get to know each other and to encourage discussions and to train social skills like doing presentations, etc. MUSSS courses are offered in other degrees like Business Informatics, Business and Economics Education, Social Economics, etc. as well. The blended learning courses have up to 1/3 face-to-face classes depending on the subject because especially tutorials or seminars including interactive elements usually require a different didactic design; therefore, learning activities are not easily reproducible virtually. For this reason, the blended learning concept is generally used in these cases. This allows students to consume parts of the course in a time and space independent manner and gives them the opportunity to deepen, practice, try out and analyze the knowledge acquired through digital learning in face-to-face courses.

Moreover, we designed a special type of MUSSS course, which is fee-based and offers a wide range of additional learning materials and special support through teachers as well as online tutors, and there is only a small number of participants during the phases of attendance. On the other hand, there are MUSSS O.C. (Open Content) courses, where the learning material is offered free of charge [5]. Besides,

[1]At the Johannes Kepler University in Upper Austria in particular, the percentage of working students adds up to 75 [3].

there are no on-campus courses, and students can choose their learning setting (time and space) completely on their own and do not have to come to the university except for the exams, which are often held in the form of electronic exams [6].

The course "Digital Business Law", which is the basis of the study on hand in this chapter, is offered as a special MUSSS version, which is described in detail below. A MUSSS O.C. type however cannot be provided since students have to interact, discuss as well as reflect in class on-campus.

With regard to our online courses, we provide a lot of different digital course material embedded into a mix of various learning environments, methods, techniques or resources like learning programs, videos, Wikis audio commented slides and microlearning tools. Furthermore, numerous tools such as forums, chats, wikis and other collaborative apps (e.g., video conferencing), technical tools or social learning activities are offered [7]. Additionally, special electronic examination rooms, where students are able to do exams on the computer, with a public key infrastructure are also provided [6].

Digital support of students' learning has become a major issue in higher education. The current learning methods give rise to new forms of (repetitive) knowledge acquisition, which improve communication, cooperation and peer interaction so that students are encouraged to collaborate and interact with each other [8].

Students can especially profit from flexibility and self-organized learning activities as well as improve their problem-solving skills [9]. Besides, self-organized learning is enhanced including all kinds of teaching and learning methods which facilitate a more self-determined learning process regarding methods, tools, study places, learning time and so forth and include teaching activities such as e-portfolio work or peer review [10].

Peer Learning

Furthermore, different pedagogical strategies are applied like case studies or peer reviews in order to enable students to profit from the benefits of collaborative learning activities or peer learning, respectively [11].

In the year 2005, the OECD was already suggesting in its dossier on "E-Learning in Tertiary Education" that universities should, among other things, promote peer learning in order to reduce the costs of e-learning [12]. Likewise, according to the European Commission, peer learning is "*… an opportunity to exchange experiences and learn from each other.*" [13]

One of the advantages of peer learning is self or peer assessment since peer learning also involves feedback, evaluation and assessment [14].

Millard, Sinclair und Newman define peer assessment or peer review as "*… a popular form of reciprocal assessment where students produce feedback, or grades, for each other's work…*" [15].

Peer review methods are well established in academic life both for research and for teaching, starting originally with art and English. In the year 1995, the University of Portsmouth reported the first computer-supported peer review learning activity. Since then, numerous forms of peer review have been in use within the scope of academic teaching [15].

So, as giving feedback to peers has been a method of quality assurance at universities for a long time, it seems useful to prepare learners for their future role as evaluators. In the English-speaking world, the teaching and learning method "evaluating classmates and peer review" is an essential part of Master's courses in particular [16]. Moreover, through peer review, students are able to practice and try out giving feedback as part of their learning process. Actually, it is an applicable approach for teachers, particularly for those educating students in the scope of Master degrees, to prepare them for their professional careers in the (near) future [17].

Additionally, peer review requires students to analyze, to evaluate and to create and, therefore, is situated at the highest level of Bloom's taxonomy of learning in the cognitive domain [18]. The learning methods outlined above can be used at different points of the learning process. Diagnostic or formative peer assessment is useful at the start as well as during the learning process, so as to be able to intervene if necessary, whereas summative assessment is applied at the end of the learning process for the purpose of grading students.

Pilot Study: Digital Business Law

In the context of this study, learners had to generate a bank of exam questions for their fellow students. In a following step, after having answered the questions of their colleagues, the learners were encouraged to provide feedback to the author of the question regarding the quality of his subject, who were then expected to improve the questions or the answers.

By evaluating the quality of contributed questions in Moodle, students provide as well as get peer feedback, and therefore, they are able to improve their self-assessment abilities. Peer review offers students a wide range of opportunities to take ownership of their learning process and to actively engage within the assessment process, hence endorsing a deeper understanding of the learning materials and the development of several skills, such as self and peer assessment, reflection and self-regulation. These skills are altogether key factors for success in the modern workplace [19].

We therefore provided the possibility to improve the student's self-regulation and reflection by inviting them to create questions for a quiz or an exam. The pedagogical strategy of having learners generate quiz questions arose for the first time in the literature in 1980 (e.g., [20]). The process of developing exam questions can be a fun, role-reversing activity that supports higher cognitive domains and the students' own learning process. Contrary to traditional teaching concepts, where students have to train, apply and memorize the course material, creating questions forces students to reflect and learn how to evaluate their acquired knowledge with regard to the course content [21]. It is thus a method which can benefit both teachers and students. Learners engaged in these activities are typically required to work harder and learn more thoroughly than through reading alone. Educators, on the other hand, are thereby able to add additional questions to their question bank [22].

9.1.1.1 eAssessment

E-learning environments use different tools and techniques for online assessment. The learning progress is controlled by using digital media for its preparation, application and follow-up. It is used for the presentation of assessment activity, the recording of responses and the administration of the tests. eAssessment is oriented to the learning objectives and is used for assessment, evaluation, documentation and feedback of learning conditions, the current level of learning and the learning outcome.

It represents the alignment of teaching, learning and assessment means, like teachers do with traditional assessment methods. eAssessment refers to the use of computer and information technology to perform the assessment process more efficiently. Generally, eAssessment should maintain the main characteristics of traditional assessment such as accessibility, validity, consistency and fairness [36, 38, 39].

Depending on the position in the learning process, various kinds of assessment are differenced, as Azevedo states: *"Formative assessment is an assessment in which constructive feedbacks are provided to a student regarding his/her knowledge and skills"* [37]. However, there is another form of assessment, called summative assessment. This is *"...the final assessment or evaluation of a student's performance, which is used to make judgments and decisions about the overall knowledge and skills of an individual"* [37]. In our case, the assessment is formative, and the students improve their learning process and prepare with peer questions for their exam.

The development of e-learning scenarios also includes models for eAssessment specifically tailored to these developments, such as the integrated model for eAssessment by Wesiak et al. [23] which has its starting point with defining the learning objectives. The various learning resources also contain different forms of evaluation, which are integrated into the assessment. Furthermore, the level of complexity can vary according to the given educational objectives as represented in Bloom's taxonomy of cognitive processes and knowledge dimensions and consequently involve questions that demand not only merely memorizing content but analyzing, evaluating and creating the content as well [18].

For each auditor or teacher, it is a challenge to generate valid exam questions that test the learning content/aims that were previously defined in the learning objectives. Objectivity and reliability are the preconditions for a valid exam, and they are the main indicators of quality standards. The technology-enhanced conducting and assessment of exams guarantees objectivity, as it is free of any subjective influence. From a practical perspective, the exam questions are also reusable, easy to administer and easy to grade. As a result of the time saved in the test evaluation, students get their feedback and test results more quickly.

Although eAssessments generally tend to test memorization rather than analytical thinking, they refer to lower levels of Bloom's taxonomy of cognitive domains [18]. It is in fact possible to construct eAssessment questions that test the students' ability to apply knowledge and to analyze problems. Grainger et al. developed a rubric based on Bloom's taxonomy for evaluating the quality of multiple choice questions. The exam questions are categorized into different levels of cognitive domains. Therefore, this framework serves as a guideline for creating exam questions that require complex cognitive thinking as well as memorization [24].

Today, eAssessment systems offer a large range of different question and answer forms, such as single choice, multiple choice, matching, yes/no or right/wrong type answers, which can be evaluated automatically. Furthermore, there are matrix questions, fill-in-the-blank text and short-text answers, calculated answers as well as the possibility to include images or animations. Different user interfaces are used like drag and drop onto text or images or content-specific question types, e.g., for accounting. Thereby, the (partly) automated correction of the exam papers plays a central role in the application of eAssessment tools.

In dealing with eAssessment systems, various dimensions have to be considered, in order to guarantee an adequately high quality. The technical solution and the infrastructure are the basis for the development of pedagogical, methodical and organizational solutions [25].

As mentioned above, the underlying intention is to guarantee that students consolidate their newly acquired knowledge by creating questions and answers themselves and additionally that they repeat the whole course content by learning and testing the exam questions provided by others. The aim was to support peer learning and to encourage students with regard to critical thinking by raising critical questions and giving useful feedback. Students should design exam questions with different levels of difficulty and cognitive processes.

9.1.1.2 Gamification Elements

New ways of learning are opening up, such as "gamification", completely new opportunities for technology-enhanced learning. Gamification describes game features that are applied in non-game context to enhance user engagement. However, since the possibilities of gamification are very diverse, e-interaction can implement a wide variety of gamification elements. In the area of education, it is an approach that tries to motivate students. The gamification element points, leaderboard and ranking are currently used on the learning platform Moodle. The use or integration of further gamification elements can be recommended, and central elements are badges, trophies, team leaderboards, levels, performance graphs, narratives or avatars.

Gamification represents an approach that tries to motivate, and it can exploit the human instinct to play, which, as a result, motivates learners or students to deal with a certain topic, to solve tasks or to discuss them with others. With gamification, game-design elements are added in the hope of incentivizing a particular process, so intrinsic motivation is added in a given gamified process which invariably uses extrinsic rewards [26]. The game elements make a potentially tedious task more fun by rewarding the user for his or her effort. This rewarding should optimally result in the user developing an effective intrinsic motivation so that he or she will execute the task without pressure in the future.

The potential of gamification is based on comprehensive motivational support and on setting flow experiences [27]. Flow as an optimal experience is characterized as a state of being fully focused and engaged in an activity. The feeling of flow is triggered by four elements: goals, rules, feedback and voluntary participation. If the difficulty of tasks is correctly balanced, it can drive the learners to a flow state which is highly motivating [28].

9.1.2 Methodology

Module Digital Business Law

In the following, the course module "Digital Business Law" is described. It is embedded in the master's program "Digital Business Management". The teaching objectives comprise in-depth knowledge of the legal framework in the decision-making areas such as data protection, domain law and computer criminal law. [29]. Students in this second-semester module have to complete a six ECTS (European Credit Transfer System) workload during the semester in the form of a seminar and a lecture. The lecture consists of 4.5 ECTS and the seminar of 1.5 ECTS.

In order to meet the needs of employed students, who presently form the majority in this master's program, an e-learning concept, consisting of an online lecture combined with an on-campus (presence) seminar, was introduced.

Through online and presence activities, interactive, social and active elements are combined, promoting self-directed and autonomous learning. At the same time, interdisciplinary competences can be acquired and students benefit from flexibility in terms of time and space.

The lecture was held as an online course (hence modeled on classic teacher-centered lessons). Through the multimedia enriched course—by means of videos, commented slides or documents—learners are guided to individual self-study and self-directed learning. Additionally, there are various forums so that students are able to ask content-related or administrative questions and also offered the opportunity to chat with the teachers. The whole course material is developed and produced in advance by the teachers and provided on our learning platform Moodle.

A seminar, however, is used for in-depth discussion and processing/working on/of practical and scientific problems using scientific methods. It is intended that students work to a great extent independently in dealing with the relevant issues and to present their findings in written and oral form [33].

Hence, since seminars usually focus on discussions and other interactive elements and therefore typically require a different didactic design, the learning activities have to be mapped in a different way when used in an online scenario. In order to avoid the restrictions that may arise when a seminar is made available online, the seminar is not held exclusively online, but two block courses (one and a half days long) are held on-campus in a typical seminar setting. So, the teacher can do in-depth discussions and students are able to do their presentations while being physically present/attending and train their practical skills face-to-face.

Previously, learners had to conduct peer learning in the form of multimedia presentations in groups for this purpose and to evaluate the work of their classmates after that. So as to increase their evaluation competency even further and to provide them with the chance to construct knowledge in an active way, in the present course, they had – in accordance with the students – to create questions, which were used – if possible – for the exam. This learning method helped them to prepare for their exam as well.

In order to enable peer and group learning, students had to form eight groups according to the chapters of the various learning topics. Afterward, each group had to generate exam questions for eAssessment, either by way of group or by individual work. They were then expected to test, comment on and rate the suggestions of their classmates. Learners could choose between various types of questions, like multiple choice questions, true/false questions, drag and drop or matching answers.

Moodle

For this reason, a plugin called "StudentQuiz" was implemented in our learning platform "Moodle" (https://moodle.org/), which allows the creation of exam questions in a multiuser mode. StudentQuiz (https://moodle.org/plugins/mod_studentquiz) is an activity in the standard Moodle installation. The version of Moodle, which is used by the students, is Moodle 3.8.1 + (Build: 20,200,228) and the "Prüfungsmoodle" (meaning "Moodle for exams"), which is a special version and only available for exams, is a Moodle 3.8.1 (Build: 20,200,113) version (Fig. 9.1).

StudentQuiz is created to support several learning aspects such as collaboration, gamification, constructivism as well as crowd-sourcing (since group ratings and comments can automize quality management) [32].

Since the students are tech savvy and they are studying "Digital Business Master", which contains many different relevant technological topics, the new Moodle feature did not raise any questions except one, which concerned a technical problem of Moodle that could be solved by our technician quickly.

The following screenshot shows one of the student-generated questions in the field of contract law. On the left-hand side, information is displayed as to whether the quiz has already been fully completed or not and the number of points which have been achieved so far. Students had to evaluate the questions and answers and to give individual comments on these questions as shown at the bottom of the page.

Students could also choose to create true–false questions as below and classmates had to test the questions and answers and to give feedback as necessary (Fig. 9.2).

The aim was therefore that the learners transfer their newly acquired knowledge into the creation of exam questions and the corresponding answers and distractors (wrong answers). Finding alternative answers that cannot be guessed or found out by eliminating wrong responses requires knowledge of the content and critical thinking. In this way, students should repeat the content of the course by learning and trying out the exam questions. The main aim was to support and facilitate peer learning and to inspire and to urge learners to raise critical questions and to give useful feedback.

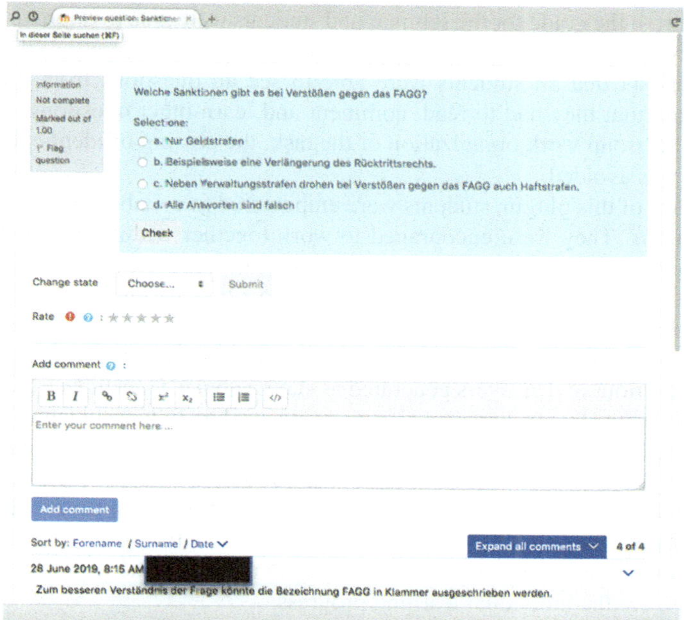

Fig. 9.1 Preview multiple choice question (Moodle)

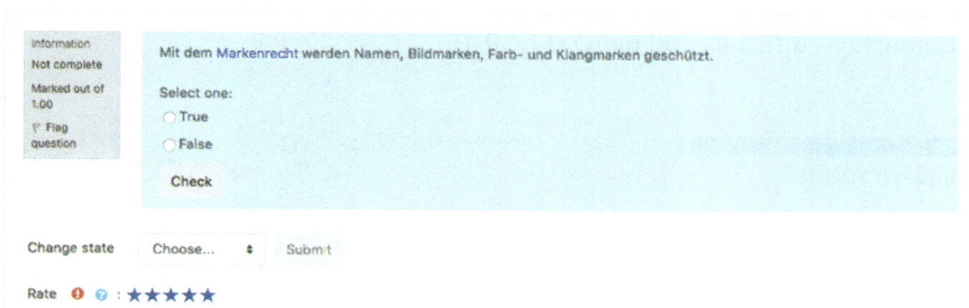

Fig. 9.2 Preview exam question (Moodle)

All relevant content was already shown in Moodle in separate chapters, and the students decided among themselves who was assigned to which topic. However, we did not check whether the entire task was carried out in the form of group work or whether students divided the questions and created them individually. Since Moodle reveals the author of a question and students were obliged to each create a certain number of questions, these questions as well as the comments were assessed. These evaluations were also included in the final grade with regard to the seminar. The whole module (lecture and seminar) is finished with a final exam, which forms the main part of the summative assessment model. The questions however were also a relevant part of the grade for the seminar and students were able to improve their final grade by one degree (the grade not sufficient excluded) with regard to the seminar. Due to the fact that all students were able to see all questions from their fellow students and that they had to read, comment and learn other questions as well, by virtue of the group work organization of the task, the posting of identical or similar questions was avoided.

By means of this plugin, students were empowered to collaboratively create their own questions. They were encouraged to work together within the question pool, prearranged by teachers or freely chosen by the learners.

Lecturers as well as learners were allowed to filter the questions into quizzes. Students were also able to complete the quizzes themselves. Another attribute of "StudentQuiz" is the option to rate the questions of their classmates and to comment on these questions and answers generated as shown above. Questions were therefore not only corrected by teachers, but also peer reviewed by fellow students. Hence, the learning opportunities for the learners were manifold: in the way of creating questions and correspondingly the respective correct and false answers, through finishing the quizzes and through commenting and rating other questions/answers. Benefits of this tool are the analysis/reporting data in the form of statistics or the ranking mode (leaderboards), which allows teachers to see the output of each student in Moodle and to compare the individual learning progress.

Naturally, students were also able to take the exam several times and to get feedback as to whether their choice was correct or not as shown here (Fig. 9.3).

The screenshot below demonstrates the appearance of such a question pool containing – since it was not anonymous – the name of the creator, title of the question, tags, difficulty and rating (Fig. 9.4).

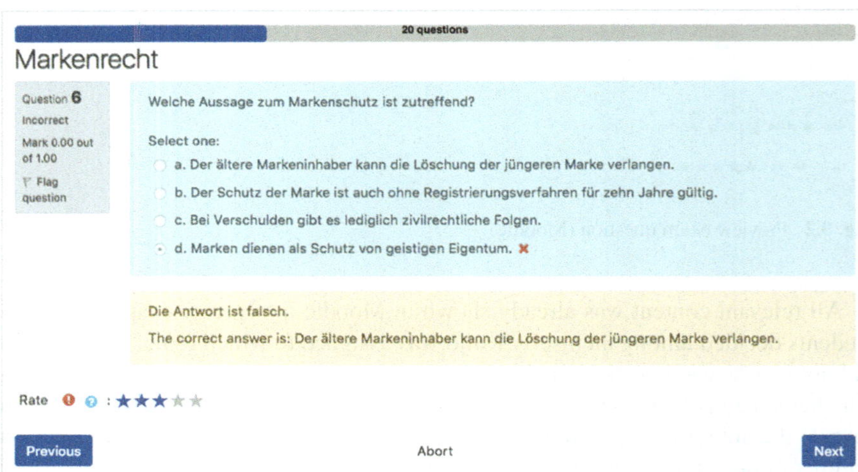

Fig. 9.3 Exams questions (Moodle)

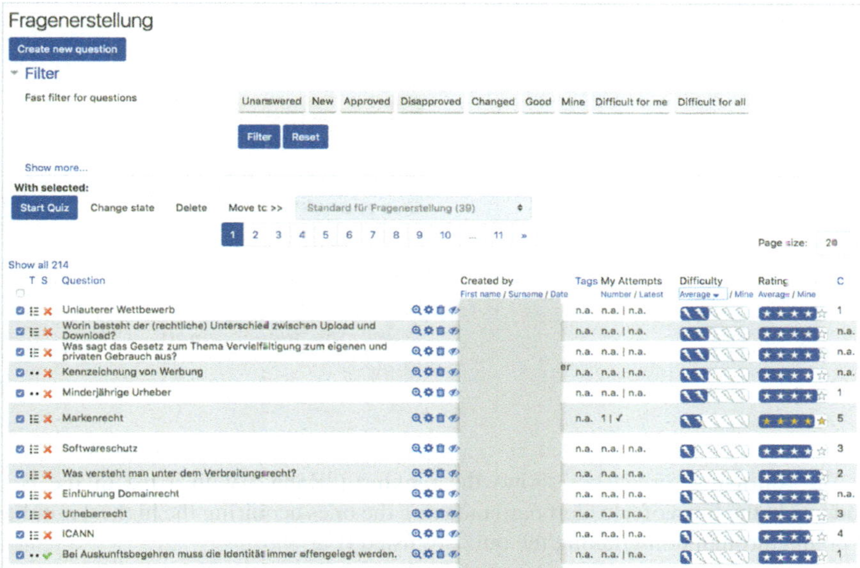

Fig. 9.4 StudentQuiz Moodle

Within the scope of the plugin, "StudentQuiz" 214 questions were produced by the participants, which means an average of seven questions and five comments per learner, whereby the commenting was compulsory too, in order to get the students to evaluate and correct the questions of their fellow students. So, even though each participant contributed only a small number of questions, a large suite was established anyway. Consequently, a huge range of questions could be generated within merely two or three semesters.

Additionally, in order to motivate students and to show their individual progress, different statistical analyses are implemented in Moodle. Firstly, there is a feature called "My progress" showing the number of correct/false attempts and the quantity of questions accepted, changed or refused (Fig. 9.5)

Secondly, gamification elements are also included in this Moodle quiz since students are able to compare each other's points or rankings. The points are calculated assigning specific points to each category of questions and are the basis for the gamification elements. So yes/no questions reach lower points than multiple choice questions or cloze. Badges can even be awarded to the student who attained the most points, which could also ensure that students are more motivated through gamification elements to create questions and to do quizzes. Increasing motivation is one of the aims of implementing gamification elements as for example Kapp [30] and Johnson [31] describe.

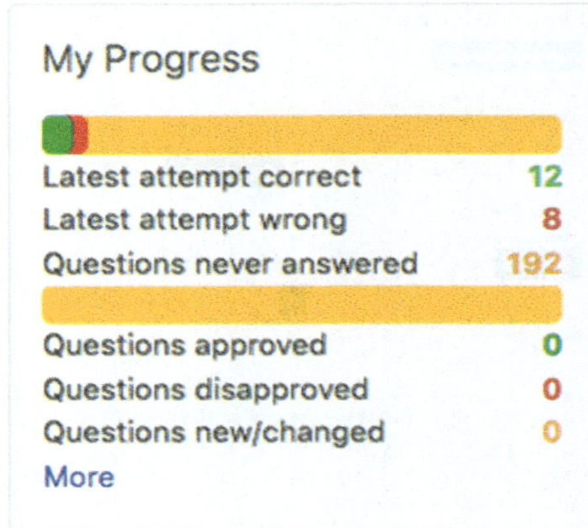

Fig. 9.5 Individual Progress (Moodle)

The following screenshot shows the ranking ("Rang") with a list of the best students in the form of a leaderboard meaning the ones acquiring the highest number of points through doing/taking the quiz are listed (Fig. 9.6).

A display (see Fig. 9.7), only available to teachers, provides a detailed list of high scores, showing total points achieved, points acquired for creation and posting of questions, points for stars (through evaluation) and for each question answered correctly.

This quantitative data is the basis for the gamification elements in the app. So the students get feedback to their ranking within the course and they can compare with their colleagues.

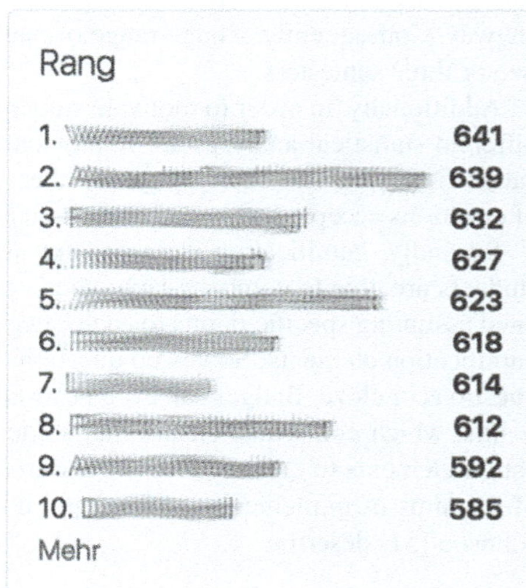

Fig. 9.6 Ranking fellow students (Moodle)

Student ranking

Rank	Fullname	Total Points	Points for questions created	Points for approved questions	Points for stars received	Points for latest correct attemps	Points for latest wrong attempts	Personal progress
1	[illegible]	640.5	90	0	126.47	424	0	100 %
2	[illegible]	638.9	90	0	124.93	424	0	100 %
3	[illegible]	631.6	90	0	126.64	418	-3	99 %
4	[illegible]	626.9	90	0	121.94	418	-3	99 %
5	[illegible]	622.9	90	0	122.86	414	-4	98 %
6	[illegible]	617.7	90	0	121.68	412	-6	97 %
7	[illegible]	614.3	80	0	110.28	424	0	100 %
8	[illegible]	611.9	70	35	97.91	414	-5	98 %
9	[illegible]	591.9	80	0	105.91	412	-6	97 %
10	[illegible]	585.3	70	35	95.29	398	-13	94 %

Fig. 9.7 Ranking participants (Moodle)

Furthermore, users are able to view the statistics concerning the community which involves those students, and optionally tutors, who are enabled to use this feature in Moodle.

Students however are only allowed to see their individual statistics, like average evaluation points, total number of created questions, the percentage of correct answers and their individual learning progress. Additionally, information with regard to the community is also displayed which should motivate students to try harder or to produce more questions for the "StudentQuiz". The information also includes the number of all created questions, the average rating of the questions, the percentage of all correctly answered questions and the average progress concerning the community of learners as is shown in the screenshot.(Fig. 9.8).

9.2 Results and Discussion

To evaluate the outcome of this pilot study and to find out whether this learning method was useful or acceptable among the students, a written questionnaire was analyzed (filled in by all students taking part in the course, which is presented in this paper). The study focuses on first experiences with peer learning in the form of creating and commenting exam questions.

The data investigated was collected by means of an online questionnaire consisting of open-ended as well as closed-ended questions. In total, 19 female and 15 male students, aged between 22 and 42 (median = 26), took part in the survey. As part of the demographic questions, students were asked whether they work while studying. This question was answered with "yes" by all of the participants except one. All

Statistics

Personal Statistics	Value	Community Statistics	Value
Number of questions you have contributed	0	Number of questions in this StudentQuiz	212
Number of your approved questions	0	Number of approved questions	14
Your received rating average	0	Rating average of all questions	4.54
Total of all your answers	27	Community average of all answers	199.37
Percentage of your correct answers	63 %	Community percentage of correct answers	95 %
Personal Progress	6 %	Average Community Progress	61 %

Fig. 9.8 Student quiz statistics (Moodle)

female students worked more than 16 h/week, ten of them more than 35 h/week, and four of the male students work less than 15 h/week.

Asked if the questions helped them to prepare for the exams 82% (see Fig. 9.9) of the participants thought that this was—at least partly—the case and nearly 90%

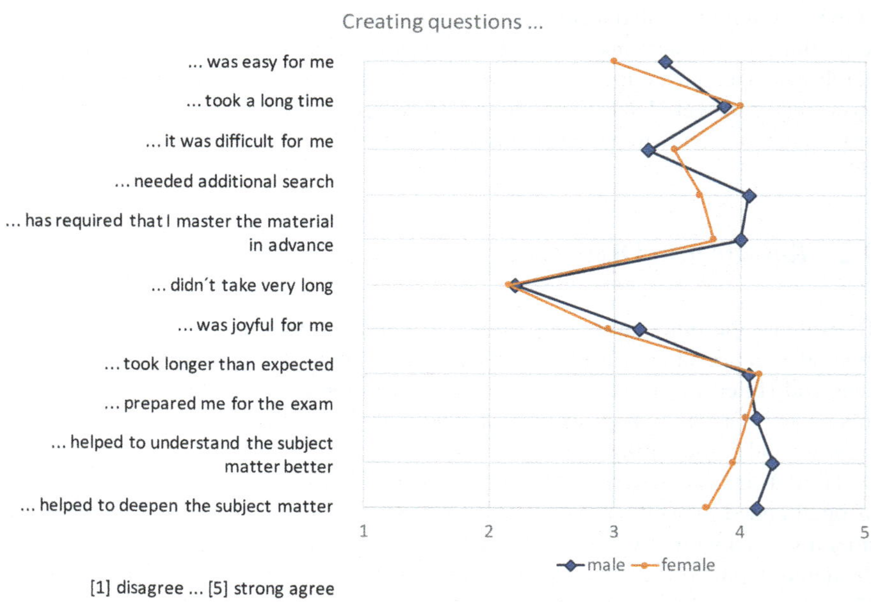

Fig. 9.9 Students rating of creating questions

noted that creating questions helped them to a better understanding of the content of the curriculum. 27 out of 34 thought that it helped them to deepen their understanding of the subject matter. In this point, a gender difference is obvious: 15 out of 15 male students and 13 out of 19 female students agreed that creating exam questions helped them to understand the subject matter better.

The students disagreed that it did not take very long to create the exam questions and agreed that it took longer than expected. In the comments, the students also noted that it was time-consuming. They rated as neutral that is was joyful for them.

The male students agreed more that the additional research and requirement to review the material enabled them to get a better and deeper understanding of the subject matter.

One of the objectives of this learning scenario was that students deal actively with the learning content. The feedback of the students shows the high involvement, but it is rather time-consuming for them – and by the way – for the lecturers, too.

Complex cognitive processes are necessary to develop exam questions. In addition to factual knowledge, conceptual knowledge is also required. In addition to remembering and understanding, analysis and evaluation of the content are also required [18]. The feedback shows that the students have dealt intensively with the subject matter.

As far as the peer learning method is concerned, a large part (85%) thought that this learning method was very good or good (see Fig. 9.10), the students rated on a four-point Likert scale from very good to not adequate. The students had a high personal learning outcome through this learning method, and they gave the highest ratings for creating exam questions to prepare for the exam, and 31 out of 34 rated

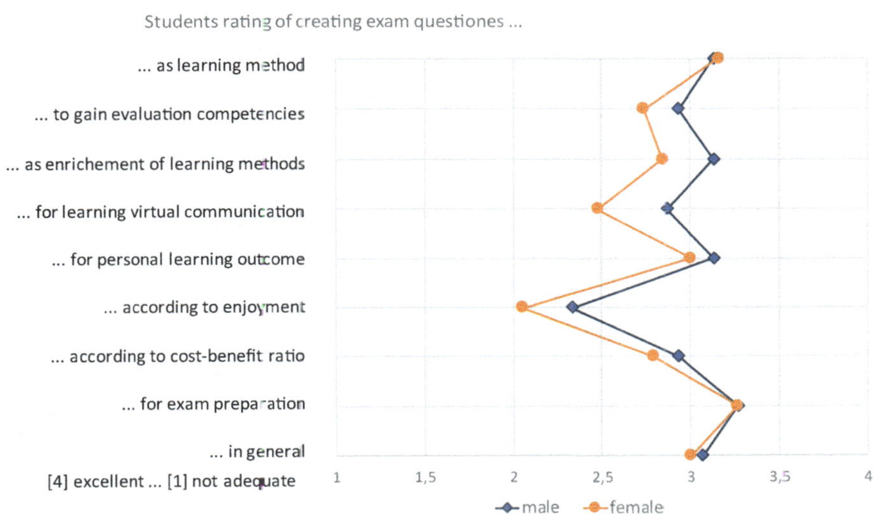

Fig. 9.10 Student perception of generating exam questions

this statement with very good or good. On average male students rated better, the enjoyment was not high.

When asked how many hours they had to put into the creating of the questions as well as the commenting, one student marked the maximum of 30 h while another one marked the minimum of 4 h. The average was 11,11 h.

It was interesting to discover that the students mainly created the questions using only the commented slides. Only 13% had additionally researched in the Internet, and not a single student had used books to create their questions. Furthermore, it was stated several times that learners would have liked more sample questions as well as more time for creating the questions and commenting on other questions. Nearly, two-thirds of the students required a list of quality criteria for the questions and example questions (see Fig. 9.13).

Figure 9.11 shows the student opinions with regard to their workflow while creating questions. The students agreed that they did in-depth analysis of the teaching material before the assessment. There is a significant gender difference in the workflow, where male students did more additional research before creating questions than the female students. Maybe they have more spare time for study than the female students who work more hours per week. They use more public Web sites for research than scientific sources.

Students on the one hand strongly agreed that the given feedback, comments and tips are helpful and constructive for the fellow students. On the other hand, the

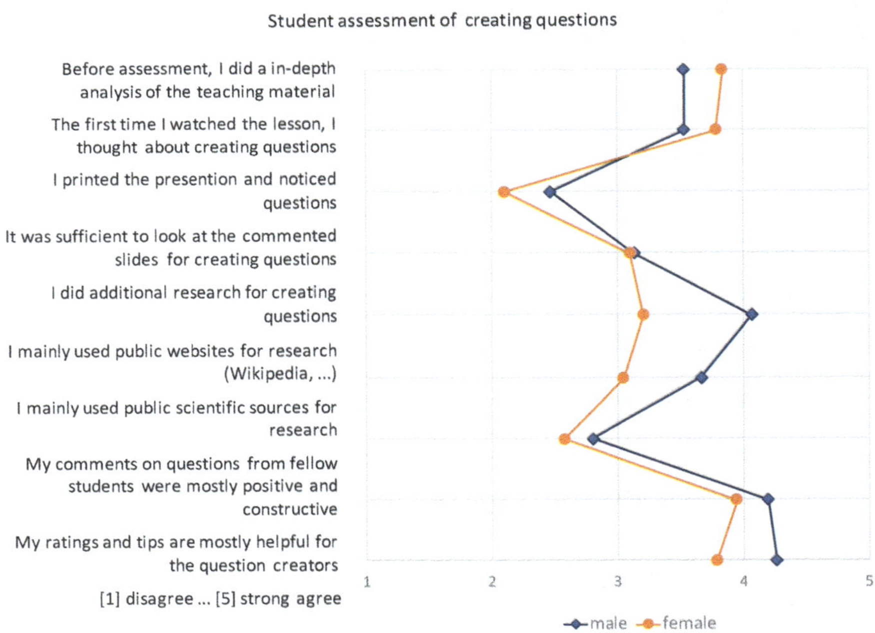

Fig. 9.11 Student´s workflow of creating questions

received feedback and comments were helpful and constructive, and this indicates a positive learning atmosphere (see Fig. 9.12).

Most of the students agreed that it was helpful for them. 83% of the students strongly or rather agreed that the feedback given was positive and constructive, and 79% found the received feedback was positive and constructive.

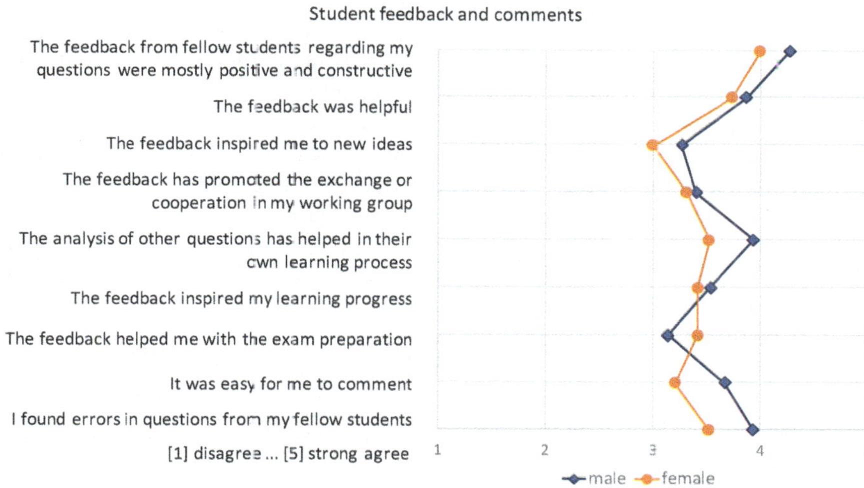

Fig. 9.12 Student feedback and comments

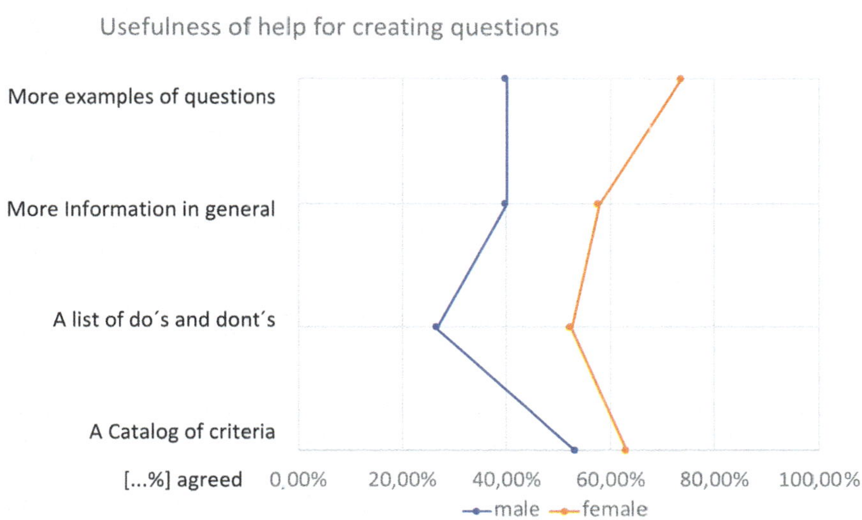

Fig. 9.13 Usefulness of more help for creating questions

On the other hand, they found errors in the questions of the fellow students. So it shows that students generating exam questions is very positive for dealing with the learning content, and it has a high impact on the learning process.

97% of the students had prepared themselves for the exam by completing the test, and 66% would have liked more cases or case studies respectively to prepare for the exam.

Two-thirds felt that the questions of their fellow students were not easy to guess and 55% thought that the questions of their fellow students were not too difficult. Almost nobody (3%) considered the questions as being not too complex at all, and only one-third did not agree that the questions went beyond the course content.

In order to provide constructive feedback and comments on the questions of fellow students, the subject matters must be analyzed and evaluated. This takes place in a self-organized learning process with the active participation of the learners. The goal of actively involving the learners in the learning scenario has thus been achieved.

A major part of the participants (90%) stated they would like to be offered other courses in the form of e-learning (in contrast to on-campus courses) as well. Furthermore, the remarks revealed that students wished for teachers to also post their comments and for more feedback loops to be provided.

Interestingly, 20% of the students printed the pdfs of the commented slides and noted questions, and in this context, only two students agreed that for creating questions it was enough to look at the slides. However, only 16% declared that they did extra research in order to create the questions.

Although 81% rated effort and use as very good or good on a four-point Likert scale, the fun factor was seen more critically. Only 35% of the participants thought that the fun factor was "good", and no one chose a "very good".

The remarks in the scope of the open questions showed clearly that respondents wished for teachers to post their comments as well and that feedback loops be provided. Furthermore, the students felt that how the questions should be generated and how difficult they should be was unclear. One student noted that he wished for: "… more information on how the questions should be created, which criteria they should meet". Another participant remarked that, "… the question-making system is helpful in learning the substance and should therefore be retained further".

The students were asked which assistance they would need for this learning activity (see Fig. 9.13). Especially the female students wanted more help and support for creating exam questions. 75% of the female students agreed that they would want more examples of questions, and about 60% of the students would like to have a list of criteria defining how to create exam questions.

Exam question generation is a new activity for students. The change in perspective, or rather the assuming of the teacher role, makes them feel uncertain, and as a result, they need some support and assistance with sample questions or a criteria catalog for best practice.

The next figure (Fig. 9.14) shows the results of the final exams in the year 2018 in comparison with the following year. In 2019, 22 students got the best marks, which could be reached (excellent), whereas one year earlier only six students gained this grade. No one got the two worst grades (four and five) in 2019.

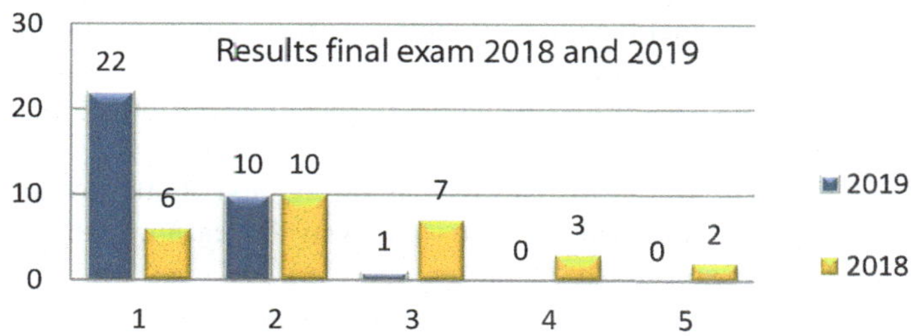

Fig. 9.14 Comparison grades of final exam

9.3 Conclusions

The aim of the learning setting that students generate and comment on e-exam questions was to activate and involve the students more in the learning process. The first experiences with this learning setting show that this objective was reached. The additional research and deepening of the learning content were a time-consuming process for the students.

To support self-directed and experience-oriented learning, peer review was integrated into learning arrangements. In order to be able to assess exam questions and provide feedback, the reviewer must have specialist expertise and must carefully examine the content of the work. This supports the learning process. However, peer review in the learning process also represents a method for quality assurance. Especially in degrees and subjects in which the previous knowledge or professional experience of the students can be used, the students are definitely critical assessors. The study has shown that students give helpful and constructive feedback.

The students´ feedback to the learning setting shows that students wish for concrete and multi-faceted guidelines in order to know how to create "good" questions. They want to know exactly what the best practices of e-learning are in advance. The majority would like to have a list of criteria and as much information as possible. In addition, sufficient time should be allowed between creating and commenting on the questions and preparing for exams based on the test questions. Many students suggested that the teacher should also write feedback in the form of a commentary.

In the next step, the time schedule will be optimized and the information concerning the creation of the questions will be revised and extended. The students will receive a catalog of criteria that helps them to create good exam questions. It is also important to point out the different levels of the cognitive processes. The number of example questions will be increased and extended with negative examples.

The tool StudentQuiz shows that it can be used as a learning scenario to activate students and to bring in new ideas. Based on Bloom's taxonomy, higher cognitive levels can be addressed when processing the learning content. With an overall high participation, the students applied quite a large number of questions to improve their learning.

It turns out, however, that the intensity of the critical approach with the learning content varies greatly. So it seems that this learning setting does not address all learners equally. This forms also a basis for further research in the field on how learners can be addressed best by this learning scenario.

Further research is necessary to investigate if the creation of questions leads to in-depth understanding and knowledge and in which way the learning can be supported by peer learning and/or peer assessment.

The described learning setting is challenging for both the learners and the lecturers. In fact, it is time-consuming, but the high involvement of the students and their learning outcomes justify the effort in this case. The limitations of this learning setting result on the one hand in the high requirements with regard to the students' expertise and self-organization and on the other hand concerning the learning content. In our case, for the master students in Digital Business Management in the field of Digital Business Law, it was successful with some limitations.

9.3.1 Future Perspectives

In the future, it will be particularly interesting to see if and where the increase in knowledge acquisition results from the preparation of the questions.

It is also planned to include more gaming elements, such as students receiving badges or rewards in Moodle for the best questions.

Furthermore, using the app, which has already been applied at the JKU and linked to Moodle for the purpose of generating the questions, is also being considered with levels and quests to promote learning through gamification.

One of the aims of the app is to offer microlearning content so that students can also learn the questions on the mobile phone. At the moment, a microlearning feature in the field of accounting is completed, which can be run in Moodle or on the mobile phone. In future, more microlearning content is planned so that students are able to train via single choice questions for theoretical exams [34]. Microlearning as a didactical concept makes learning easier by frequent change of activities to be learned and employs microcontent as a foundation for knowledge building. The learning content is broken down into small units and short activities to use time flexibly [35]. The high number of generated e-exam questions enables on the one hand the use of gamification elements. On the other hand, the learning process can be analyzed to give students feedback on their learning progress. In the expected future, it is planned to use artificial intelligence in the form of learning analytics tools in order to create a learning path, which is individually adapted for each student and to evaluate the creation of the questions in depth. Furthermore, customized and personalized online learning as well as real-time questioning and the determination of the learner's behavior is planned in future.

References

1. Zaussinger, S., Unger, M., Thaler, B., Dibiasi, A., Grabher, A., & Terzieva, B. (2016). *Projektbericht Studierenden-Sozialerhebung 2015*. Institut für Höhere Studien (IHS): Bericht zur sozialen Lage der Studierenden.
2. Schlageter, G., Feldmann, B. (2002). E-Learning im Hochschulbereich: der Weg zu lernerzentrierten Bildungssystemen. In: L.J. Issing, P. Klimsa (Eds.), *Information und Lernen mit Multimedia und Internet. Lehrbuch für Studium und Praxis*, 3th edn. Beltz, Weinheim, pp S 347–357
3. Steinbock, H. (2015). Was JKU-Studenten verbessern würden, Oberösterreichische Nachrichten. https://www.nachrichten.at/nachrichten/politik/landespolitik/Was-JKU-Studenten-verbessern-wuerden;art383,1723838. May, 5 2015.
4. Katzlinger-Felhofer, E., Windischbauer, U. (2010). Multimedia Study Services–A Blended Learning Approach for Part-time Bachelor Students in the Study Field of Economics, Business or Social Sciences. In A. Szucs, A.W. Tait (Eds.), *EDEN Annual Conference 2010* Valencia, Spain, pp. 493–500.
5. JKU. (2018). MuSSS. https://www.jku.at/studium/studienarten/multimedia-fernstudien/musss/. January 18, 2020.
6. Katzlinger, E., Höller, J. (2016). Public key infrastructure for e-assessment. In Y. Li, M. Chang, M. Kravcik, et al. (Eds.), *State-of-the-Art and Future Directions of Smart Learning*. Springer Singapore, Singapore, pp 287–291. doi:https://doi.org/10.1007/978-981-287-868-7_34.
7. Niederländer, U., Katzlinger, E. (2018). Supporting virtual learning for digital literacy. First experiences with a mobile app and gamification elements.
8. Ge, Z.-g. (2011). Exploring e-learners' perceptions of net-based peer-reviewed English writing. *International Journal of Computer-Supported Collaborative Learning, 6*(1), 75–91.
9. Ehlers, U.-D. (2010). Qualität für digitale Lernwelten: Von der Kontrolle zur Partizipation und Reflexion. In K.-U. Hugger & M. Walber (Eds.), *Digitale Lernwelten: Konzepte, Beispiele und Perspektiven* (pp. 59–73). Wiesbaden: VS Verlag für Sozialwissenschaften.
10. Katzlinger, E., Herzog, M.A. (2014). Intercultural Collaborative Learning Scenarios in E-Business Education: Media Competencies for. Multicultural Awareness and Technology in Higher Education: Global Perspectives: Global Perspectives, pp. 24–46.
11. Bauer, C., Figl, K., Derntl, M., Beran, P. P., & Kabicher, S. (2009). Der Einsatz von Online-Peer-Reviews als kollaborative Lernform. *Wirtschaftsinformatik, 2*(2009), 421–430.
12. OECD. (2005). Policy Brief-E-learning in Tertiary Education, OECD. https://www.oecd.org/edu/ceri/35991871.pdf. March, 16 2016.
13. Europäische Kommission. (2014). Modernisation of Higher Education-Report to the European Commission on new modes of learning and teaching in higher education. https://ec.europa.eu/dgs/education_culture/repository/education/library/reports/modernisation-universities_en.pdf. November, 9, 2016
14. Boud. (2001). Introduction: making the move to peer learning. In: Boud, Cohen und Sampson (Eds.), *Peer learning in higher education learning from & with each other*. p. 3.
15. Millard. (2008). Peer Pigeon: A Web Application to Support Generalised Peer Review, in: *E-Learn 2008-World Conference on E-Learning in Corporate, Government, Healthcare, and Higher Education*, p. 3824.
16. Hoidn, S., Kärkkäinen, K. (2014). *Promoting Skills for Innovation in Higher Education.*
17. Brill, J.M. (2016). Investigating peer review as a systemic pedagogy for developing the design knowledge, skills, and dispositions of novice instructional design students. *Educational Technology Research and Development*, pp. 1–25.
18. Anderson, L.W., Krathwohl, D.R., Bloom, B.S. (2001). A taxonomy for learning, teaching, and assessing: A revision of Bloom's taxonomy of educational objectives. Allyn & Bacon.
19. Kay, A. E., Hardy, J., & Galloway, R. K. (2018). Learning from peer feedback on student-generated multiple choice questions: Views of introductory physics students. *Physical Review Physics Education Research, 14*(1), 010119.

20. Rakes, S. K., & Smith, L. J. (1987). Strengthening comprehension and recall through the principle of recitation. *Journal of Reading, 31*(3), 260–263.
21. Jones, J. A. (2019). Scaffolding self-regulated learning through student-generated quizzes. *Active Learning in Higher Education, 20*(2), 115–126.
22. Filip, A., Pudło, W., Marchewka, D. (2018). Innovative Learning: Students in the Process of Exam Quizzes Building. Blended and Online Learning, p. 184.
23. Wesiak, G., & AL-Smadi M, Höfler M, Gütl C, . (2013). Assessment for complex learning resources: Development and validation of an integrated model. *International Journal of Emerging Technologies in Learning (iJET), 8*(S1), 52–61.
24. Grainger, R., Osborne, E., Dai, W., & Kenwright, D. (2018). The process of developing a rubric to assess the cognitive complexity of student-generated multiple choice questions in medical education. *the Asia Pacific Scholar, 3*(2), 19–24.
25. Gruttmann, S. (2010). *Formatives E-Assessment in der Hochschullehre.* Computerunterstützte Lernfortschrittskontrollen im Informatikstudium: Westfälische Wilhelms-Universität Münster, Münster.
26. Urh, M., Vukovic, G., Jereb, E., & Pintar, R. (2015). The Model for Introduction of Gamification into E-learning in Higher Education. *Procedia-Social and Behavioral Sciences, 197,* 388–397. https://doi.org/10.1016/j.sbspro.2015.07.154
27. Blohm, I., & Leimeister, J. M. (2013). Gamification. Business & Information. *Systems Engineering, 5*(4), 275–278. https://doi.org/10.1007/s12599-013-0273-5
28. Wiggins, B. E. (2016). An Overview and Study on the Use of Games, Simulations, and Gamification in Higher Education. *International Journal of Game-Based Learning, 6*(1), 18–29. https://doi.org/10.4018/ijgbl.2016010102
29. JKU. (n.d.). Studienhandbuch. https://studienhandbuch.jku.at/101728. January, 7 2020.
30. Kapp, K. M. (2012). The Gamification of Learning and Instruction: Game-based Methods and Strategies for Training and Education. pp. 9–275.
31. Johnson, L., Becker, S., Cummins, M., Estrada, V., Freeman, A., & Ludgate, H. (2013). *NMC Horizon Report: 2013 Higher* (Education). Austin, Texas: The New Media Consortium.
32. Moodle. (2019). StudentQuiz. https://docs.moodle.org/38/de/StudentQuiz#Einf.C3.BChrung. May 3, 2020.
33. JKU. (2019). Satzung der Johannes Kepler Universität Linz. Mitteilungsblatt vom 19.06.2019, 31. Stk., Pkt. 431. pp. 11
34. Katzlinger, E., Niederländer, U. (2018). Supporting Virtual Learning for Digital Literacy: First Experiences With a Mobile app and Gamification Elements. In: Ntalianis K, Andreatos A, Sgouropoulou C (2018). Proceedings of the 17th European Conference on e-Learning ECEL 2018. ACPI. pp. 235–244
35. Behringer, R. (2013). Interoperability Standards for MicroLearning. International MicroLearning Conference 7.0, Stift Goettweig pp. 1–10.
36. Shute, V. (2009). Simply Assessment. *International Journal of Learning and Media, 1* (2), 1–11. https://doi.org/10.1162/ijlm.2009.0014.
37. Azevedo A, Azevedo, J (2018) Handbook of Research on E-Assessment in Higher Education. IGI Global: Hershey PA p 29
38. Baker, E. L., O'Neil, H. F., & Linn, R. L. (1993). Policy and validity prospects for performance-based assessment. *American Psychologist, 48*(12), 1210–1218.
39. JH McMillan 2013 Research on Classroom Assessment Sage Publications Thousand Oaks

The manufacturer's authorised representative in the EU is Springer Nature Customer Service Centre GmbH, Europaplatz 3, 69115 Heidelberg, Germany. If you have any concerns regarding our products, please contact ProductSafety@springernature.com

Printed and bound by CPI Group (UK) Ltd, Croydon, CR0 4YY

30/09/2026

02225287-0007